THE
GENIUS ENGINE

WHERE MEMORY, REASON, PASSION, VIOLENCE, AND CREATIVITY INTERSECT IN THE HUMAN BRAIN

KATHLEEN STEIN

John Wiley & Sons, Inc.

Published by John Wiley & Sons, Inc., Hoboken, New Jersey
Published simultaneously in Canada

Illustration credits: Page xi, copyright Christoff Lab, adapted with permission from christofflab.psych.ubc.ca; page xii, image on lower left courtesy of McGill University, www.thebrain.mcgill.ca; page xiii, Christoff et al. (2001) *NeuroImage, 14,* 1136–1149.

Design and composition by Navta Associates, Inc.

For general information about our other products and services, please contact our Customer Care Department within the United States at (800) 762-2974, outside the United States at (317) 572-3993 or fax (317) 572-4002.

Wiley also publishes its books in a variety of electronic formats. Some content that appears in print may not be available in electronic books. For more information about Wiley products, visit our web site at www.wiley.com.

ISBN 978-0-471-26239-8

Printed in the United States of America

For Margaret Bruce Stein

Contents

Acknowledgments

This project became real to me a few months after September 11, 2001, at the Ontario–New York State border, crawling through an interminable line at customs. I had recently left the lab of Vinod Goel at York University in Toronto, and was now mentally replaying Goel's accounts of the prefrontal cortex's wondrous reasoning and problem-solving capabilities. While waiting as cars, buses, and trucks were searched for bombs, I thought how, more than ever, we need these special functions of the prefrontal cortex.

In this endeavor, I had the good fortune to draw upon the wisdom, enthusiasm, and generosity of a number of superb neuroscientists, including Jonathan Cohen, Joaquin Fuster, Earl Miller, Jeremy Gray, Todd Braver, Randy O'Reilly, Brenda Milner, Michael Petrides, Kent Berridge, Adrian Owen, Steven Anderson, David Zald, Josh Greene, John Gabrieli, Mario Beauregard, Andrew Conway, Stephen Goldinger, Arthur Shimamura, Russell Epstein, Petr Janata, Birgit Mathiesen, and Bruce Price. Special thanks to forensic psychiatrist Monty Brower for his passion and empathy for those adrift. And to Kalina Christoff, whose great discoveries are just beginning.

I'd also like to thank neuroscientists exploring the evolutionary origins of the frontal lobes, whose work is a subtext here—Jonathan Kaas, Todd Preuss, Leah Krubitzer, and Barbara Finlay—and to salute those engaged in extraordinary efforts to understand the disorders of the prefrontal cortex, including E. G. "Ted" Jones, Wayne Drevits, Hilary Blumberg, and David A. Lewis. And to Adele Diamond and Elizabeth Sowell, for their investigations of the developing brain.

Many thanks also to the wise Dick Teresi for pointing me in the right direction, Randy Long for design, my agent Michael Carlisle, and Douglas Stein for his precise and seemingly effortless grasp of brain anatomy and physiology. Thanks, finally, to my editor, Stephen S. Power, for seeing the big picture.

Brain Maps and Matrices Diagram

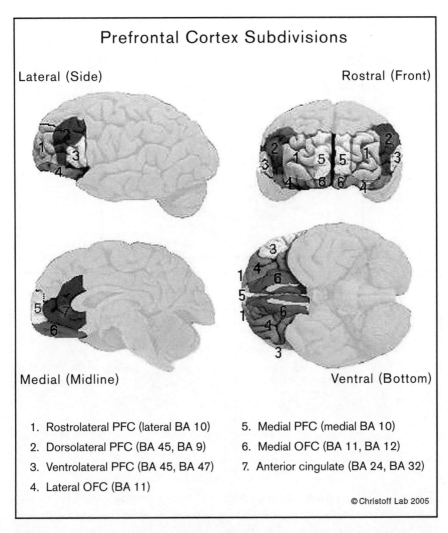

Prefrontal Cortex Subdivisions

Lateral (Side)

Rostral (Front)

Medial (Midline)

Ventral (Bottom)

1. Rostrolateral PFC (lateral BA 10)
2. Dorsolateral PFC (BA 45, BA 9)
3. Ventrolateral PFC (BA 45, BA 47)
4. Lateral OFC (BA 11)

5. Medial PFC (medial BA 10)
6. Medial OFC (BA 11, BA 12)
7. Anterior cingulate (BA 24, BA 32)

© Christoff Lab 2005

Figure 1. These are the several neighborhoods of the human prefrontal cortex as seen from four points of view. This diagram reflects some recent findings from brain imaging regarding specialization of prefrontal subdivisions. It is the most complete picture to date of PFC subdivisions in the frontal lobes.

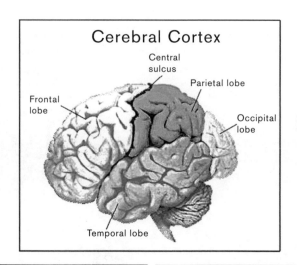

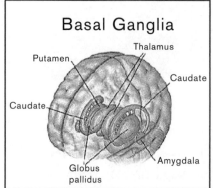

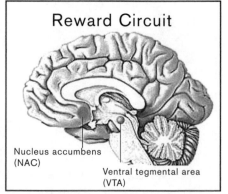

Figure 2. Top diagram: the four lobes of the cerebral cortex. The images below are schematic diagrams of important subcortical networks. At left: the basal ganglia, involved in the regulation and coordination of movement generated by the cortex and implicated in the flow of thought. At right: a reward circuit involved in the experience of pleasure and implicated in addictions.

Raven's Progressive Matrices

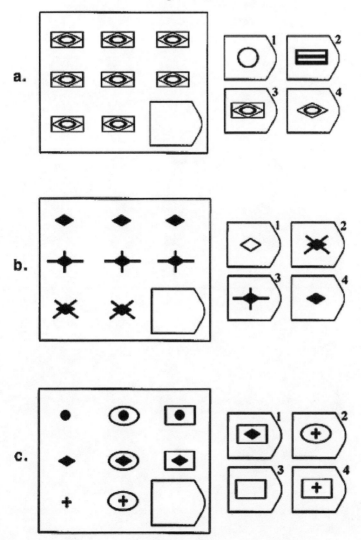

Figure 3. Raven's Progressive Matrices present problems at different levels of complexity. Participants have to infer the missing figure and select it from the four choices in the boxes at right. The matrix at top (a) is the simplest, "0-relational," problem; the matrix at middle (b) is a slightly more difficult 1-relational problem; and the matrix at bottom (c) is a 2-relational problem and is the hardest, since you must simultaneously consider both horizontal and vertical relationships. Correct answers for the example given here are (a) 3, (b) 2, and (c) 4. These are easy examples. Raven's Matrices can be diabolically difficult.

INTRODUCTION

> The Frontal Lobe, despite decades of intensive research by physiologists, anatomists and clinicians, has remained the most mystifying of the major subdivisions of the cerebral cortex. . . . no entirely satisfactory hypothesis concerning the neural mechanisms of the frontal lobe would seem attainable at present.
>
> —*Walle Nauta,* "The Problem of the Frontal Lobe: A Reinterpretation" *(1971)*

It was at the seething, human-hive-like Society for Neuroscience meetings of the late 1980s that presentations by Yale's Patricia Goldman-Rakic on the prefrontal cortex began to transfix huge audiences. Her talks and slide shows hinted at palpable, multidimensional answers to the ancient mysteries of higher mental functions. In one groundbreaking study after another, Goldman-Rakic began to elucidate how little-understood prefrontal operations generate the mindprint of our individuality, sense of

1

identity, ability to base action on knowledge, and yes, perhaps, even the evolution of consciousness itself.

This foremost part of the cerebrum, the prefrontal cortex, is the twenty-first-century brain region. Because we are immersed in an ever-accelerating data flow, our brains are being called upon to process and prioritize ever more information, ever more quickly and simultaneously. Electronic gear serves as adjunct plug-in cortical units, urging our minds to hurry up, attend to, and schematize our multiple ongoing scenarios. Multitasking is the default mental setting now. A distinct set of prefrontal processes, argues the British neuropsychologist Paul Burgess, may be essential to multitasking, which, he claims, is "at the very heart of competency to everyday life."[1]

Prefrontal functions are being legislated for and against daily. Laws banning motorists from talking on their handheld phones while driving judge that most humans lack the capacity to adequately process a river of driving information while fiddling with and conversing on a phone. Whether we can effectively juggle multiple data streams, make judgments, and act upon them more or less simultaneously is a larger issue. Although many of us can hold at least six information chunks in mind for seconds, we are unaware of how easily active memories can be swept away when we focus on one thing or are distracted by a problem we are hard-pressed to solve. Since there are limits to how much data we can hold in our working memory banks, we often are blind to important things that pass right before our eyes.

Most people are still fairly unknowing about this brain region that is at the core of our humanity. When someone mentions the prefrontal cortex (PFC), it is still most often in the context of lobotomy—either as an edgy joke or to connote another person's feeble intellectual balance sheet. Indeed, fifty years ago in the United States, perhaps only the atomic bomb elicited more dread and loathing than lobotomy. Just the mention of it could produce a gothic shiver at the thought of depersonalization at the twist of a leucotome. A shadowy account of a troublesome relative taken away and, with the upward jab of an ice pick through the top of the eye socket, transformed into a zombie—that would silence a roomful of people or straighten out a rebellious teenager.

For those who studied brain structure and function back then, lobotomy provided only marginal insight into the PFC. The results of the surgery varied enormously from patient to patient. Some experienced little apparent change; others were reduced to walking vegetables. There was little systematic follow-up or documented study of individuals before and after lobotomy. About the prefrontal cortex, neurobiologists of the twentieth century were basically clueless. The prefrontal cortices were still the "silent lobes."

Thus the singular occasion that possibly sounds the opening bell of modern exploration of the prefrontal cortex is marked by a certain irony. It was August 1935, at the Second International Neurological Conference in London, and Carlyle Jacobsen, a young experimental psychologist from Yale, was delivering a paper to an audience that included his mentor, the distinguished neurophysiologist John F. Fulton, and the Portuguese neurologist Antonio Egas Moniz, later dubbed the "father of lobotomy." Before a full house, Jacobsen described a series of experiments on two chimpanzees that conclusively demonstrated, some of his peers felt, that the frontal lobes of the primate are the brain region responsible for "the ultimate adjustment of the individual to the environment."[2]

One of Jacobsen's experiments, called the "delayed-response test," probed monkeys' short-term memory. It was an ingeniously simple experiment, a kind of two-card monte, in which an animal watched as his trainer put a food reward under one of two cups. Then the handler lowered a screen, blocking the reward from view. After varying intervals of time, the screen was raised and the monkey could choose the cup and reward. To win the reward, the animal had to remember where he'd last seen the treat. He had to "hold" this memory in his mind after its object had vanished from view. After Jacobsen had surgically removed a portion of the monkeys' prefrontal cortex, they could no longer remember which cup hid the treat after only a few seconds' delay. This short-term deficit, Jacobsen proposed, showed that the prefrontal cortex was necessary for an individual's intelligent interaction with the world around him or her. This prefrontal memory capacity demonstrated a basic and remarkable function of the prefrontal cortex.

Jacobsen next conducted the same tests on two chimpanzees. Although Lucy was fairly laid-back, Becky sometimes grew violent when frustrated. Her errors in the experimental tasks triggered tantrums during which she flew into a rage, rolled on the floor, and defecated. After her prefrontal surgery, however, Becky was no longer volatile but seemed to take on a certain "c'est la vie" attitude. It was as if, quipped Jacobsen, "the animal had joined the happiness cult of the Elder Micheaux and had placed its burdens with the Lord."[3]

Jacobsen tossed off the wisecrack to lighten his technical account of lab protocol. But many in the audience, fixating on the spin-off finding—that removing the prefrontal section had calmed the savage beast—had ears only for the joke. Moniz and others soon began experimenting with surgical severance or removal of prefrontal sections in humans with intractable mental disorders. They reported some success in the operations.

Enthralled by Jacobsen's accidental finding, Fulton himself performed more operations on chimps, and hyped Becky's emotional transformation to the scientific community. Becky's story became the apparent rationale for Fulton to experiment with the surgery on patients with severe psychotic symptoms. But as aficionados of lobotomy history know, the procedure was wrested out of the lab and into the asylums by the neurosurgeons Walter Freeman and James Watts, who went on to do thousands of lobotomies.

To Jacobsen, Becky's emotional taming was an experimental sidebar. Feeling no need to link a brain surgery that quelled neurotic behavior in a chimp to human psychosis, he pursued the phenomenon no further. Carlyle Jacobsen's central discovery—that the prefrontal cortex is essential for flexible short-term memory operations—was, for a time, buried beneath the lobotomy juggernaut. So Jacobsen is scarcely known outside the neuroscientific community, and to the extent he is, it is for something he didn't deign to study. But decades later, when instruments for studying the silent lobes were orders of magnitude more sophisticated, Jacobsen's version of the delayed-response test would become a vital research model in human as well as animal studies. It is ironic, too, that the term "prefrontal cortex," first employed in 1868, didn't come back until the 1935 Becky paper where Jacobsen reintroduced it.

Despite Jacobsen's findings, then, for two-thirds of the twentieth century the prefrontal cortex did remain silent. A few visionary adventurers evaluated victims of frontal lobe strokes or tumors, or, more often, the mutilated brains of soldiers in the two world wars and Vietnam. Military combat, far more than anything else, advanced knowledge of the frontal lobes. By the end of the American

Civil War, muskets had given way to rifles with bullets, creating smaller head wounds and thus more discrete behavioral deficits. From the human abattoirs of World War I, doctors at combat hospitals saw an unprecedented number of men with chunks of their frontal lobes blown away. Just how much battering a frontal lobe could absorb before noticeable behavioral change took place remained unclear. The neurologist Kurt Goldstein in Germany, and later Alexander Romanovich Luria in Russia, sought to understand the frontal lobes by linking brain trauma to behavioral abnormalities. After World War II, the Soviet Union had more head-wounded troops than any other nation, so Luria could test his growing ideas about brain structure and function on a tragic abundance of cases. In the 1950s and 1960s, Brenda Milner in Canada redesigned Jacobsen's animal experiments for study of humans with PFC damage, shedding new light on the subtle workings of this mysterious region.

But few scientists dared to go there. The PFC was too intractable, too weird. And that was the way things stood until almost the end of the twentieth century. Certainly, research would be transformed by new instrumentation—such as PET and MRI imaging—but one cannot ignore the influence of two very different personalities who inspired others to venture into the uncharted prefrontal zones.

I first became fascinated by the prefrontal cortex at conferences where Goldman-Rakic delivered riveting presentations on the unique capabilities of prefrontal neurons. The petite, fierce neurobiologist was exploiting fairly new microelectrode technology to tap into multicellular crosstalk among the PFC's brain cells; she was eavesdropping on the voices of silent lobes. Goldman-Rakic worked with a zeal and thoroughness like few others. Her enthusiasm for the wonders of the PFC was infectious. She took Jacobsen's discovery of the delayed-response phenomenon and gave it a name: working memory. She saw that this PFC-based capacity to hold an image "in mind" is perhaps the brain's most flexible mechanism, and its evolutionarily most significant achievement. "Here is this machine that can hold on to information after that information disappears from view, or out of hearing range," she declared. "That holding operation is what connects Time One to Time Two. That holding operation is the glue of our conscious experience."

Another pioneer, the UCLA neurobiologist Joaquin Fuster,

preceded Goldman-Rakic by over a decade in spying on the prefrontal cortex. Fuster had perceived that certain PFC neurons had
unique and remarkable polymorphic talents. These brain cells were
exuberant multitaskers capable of recombining as a single neural
"byte" diverse information in myriad configurations. Fuster saw the
PFC's remarkable ability to hold online dynamic, complex mental
"movies" over time past to time future. Under Fuster and Goldman-
Rakic's intense scrutiny of these special memory cells, the PFC was
beginning to speak.

Yet even now the prefrontal cortex remains the last great wild
place in the brain. This neural territory that, more than any other,
defines what it is to be human, is slow to give up its secrets. But
beginning in the 1990s, advanced imagining technology, allowing
for noninvasive observation of the human in action, began to open
unprecedented opportunities for studying the neural bases of
"thinking" in higher cortical areas. Today, instead of a small cadre
of PFC researchers, there are hundreds. And no brain region is
quite so in fashion as the human prefrontal cortex.

Although other primates, and indeed many other animals, have
working memory, in no other species is this prefrontal function so
highly articulated, so greatly enhanced, as in humans. The story
of the prefrontal cortex, then, is the narrative of the species *Homo
sapiens*—our prevailing spirit, distinctive character, talent, aptitude,
and inclination. Our genius.

The prefrontal cortex is the agent of mental suppleness. It can
evoke a moment in remembered time and hold together the mental representation of one's personal narrative of past, present, and
the "remembered future" we may call "foresight." The PFC enables
a person to act today in light of past actions, and to predict how he
or she will act tomorrow in light of an anticipated set of conditions
that constitute "reality." The PFC is what allows us to gather
together and retain online all the elements we need to create our
internal film scripts, while we transform them into dynamic works
of strategy and art. Think of the dictionary synonyms for "creative": artistic, clever, deviceful, fertile, formative, gifted, hep, hip,
ingenious, innovational, innovative, inspired, original, originative,
productive, prolific, stimulating, visionary, way out. The PFC does
not provide all the information processing in these acts of invention,
but it is central to making them happen.

To chart the PFC is to tell the story of how we humans have progressively controlled our surroundings less and less through direct physical acts, and more and more by strategic choices, internalized mental constructs, ideas, plans, and their externalized manifestations. These are the tools, language, and technologies—the appointments of civilization, science, and culture. To a great extent, these are the fruits of the prefrontal cortex and its far-flung neural networks. The PFC is said to be the headquarters of "executive function," a catchall term with evocations of CEOs, top cats, or the Oval Office. The term "executive" now also reverberates with other meanings, from Enron wrongdoings to the exe. command ubiquitous in computer operating systems. "Executive function" in the brain, too, is a tricky concept and devilishly hard to define.

So how does the PFC serve as the central executive, the in-house "genius"? Lingering in philosophy, psychology, and brain research has been the nagging problem of "who is in charge." Often, experts have used the device of a homunculus (Latin's "little man") to explain an unknowable prime agent of the mind; they employed an internalized creature who crouches up there in the brain and experiences the world through the eyes, ears, nose, and touch; who cultivates thoughts and plans; and who executes willful actions. The homunculus has been stubbornly entrenched as a central construct in numerous theories of consciousness and identity. As mental function has become better understood, this iconic gnome has been pushed farther and farther "up" into the brain, into the lesser-charted areas of the "association cortices," until it seemed that Mr. Homunculus was making his last stand in the prefrontal cortex. But no matter where the little man was enthroned, there existed an intrinsic problem with the homunculus thingamabob, logically, and in terms of brain structure and function.

Although it is sometimes hard to dispel the illusion that there is a daemon in one's brain whose function it is to execute "me" functions, the homunculus cannot be a true explanation for one's self or will. The homunculus represents but a repetition of the original "who's in charge?" question on another level. The philosopher Gilbert Ryle in 1949 argued that if there were a little agent inside guiding the brain's thoughts, then would not the homunculus "himself" require another homunculus, ad infinitum, in an infinite regress? It is biologically contradictory, furthermore, to cede my ability to "choose" to the homunculus, since the homunculus's agency is separate, other than "me." How can this micro "not me" be the executor of my macro will?

The homunculus as muse was nonetheless deployed as a misguided deus ex machina to explain higher brain function for lack of satisfactory answers to the who's in charge question. But discovery of how the PFC works and talks to the rest of the brain is now helping us to banish the little Wizard of Oz. As we will see, it is a much more interrelational, reciprocal brain we have. Prefrontal cortex investigations are revealing a homunculus in the process of being reconfigured into a dynamically interdependent web, an internalized ecosystem of the self. The true genius of the PFC lies in its interconnections, feedback, feedforward, loops within loops. What is so exciting is to see how this genius of the PFC emerges not only from what it controls, shapes, and curtails but from what it listens to and depends on in the rest of the brain, as we willfully connect with our environment.

The Unique Tissue

At the core of all the creativity and logic in the PFC is the simplest building block of cognition: working memory. It is the most understood function of the prefrontal cortex, its operational epicenter located directly to the side of each eyebrow in the dorsolateral PFC (see figure 1 on page xi). Anatomical loci, such as the dorsolateral PFC, are generally called Brodmann areas (BAs), named for the fin de siecle German anatomist Korbinian Brodmann. Let's give Brodmann credit here: he was the cartographer of the cortex who distinguished the prefrontal regions within it.

Despite the ubiquity of his name, Brodmann is little celebrated as a man and visionary scientist; there are no handy biographies, no annual Festschrifts. Yet always viewing cortical topography as a functional unity, Brodmann sought to understand how its various parts existed within the whole, eventually mapping the entire mammalian cerebral cortex. Within the human cortex he numerically designated over fifty areas, each with its own distinctive cell type, density, and other architectonic features. Some areas he further subdivided. This endeavor culminated with the 1909 publication of *Vergleichende Lokalisationslehre der Grosshirnrinde in ihren Prinzipien dargestellt auf Grund des Zellenbaues*, translated as *Localisation in the Cerebral Cortex*. The text still stands as the chart for cortical navigation. Despite flaws, "Brodmann" is the most commonly used reference in neuroanatomy. To refer to a specific cortical region, most everyone to this day cites him.

Brodmann's classification system endures in part because it is an elegant articulation of brain design that transcends his place in history. As a captive of nineteenth-century thinking, he came from a background replete with Hegelian ideas passé even then. And raised in a milieu emphasizing phylogenetic scale, logically he should have opted to line up the brain structures of mammals progressively. But he avoided that. Instead, he explored the possibility that different species evolved their own unique specializations. Brodmann showed how it's possible to work within a particular framework without being straitjacketed by it: the sign of a remarkable artist or scientist. During his life he was given meager encouragement. Ironically, his grant applications were routinely rejected with comments like, "None of this work is of any lasting value." For famous bad reviews Brodmann was right up there. Irascible and difficult to work with, he died in 1918 at age forty-nine, possibly from an infection contracted while doing a research autopsy.

From preliminary sketches one imagines how with a single stroke Brodmann drew a thick black line to bisect the cortex's four lobes into front-to-back sections. This line is the massive crevasse of the central sulcus, the huge rift that seems to nearly cleave each cortical hemisphere crosswise. Everything seems to radiate from it forward or backward. Everything in the cortex behind the central sulcus is about receiving, storing, perceiving, integrating, and interpreting information; everything in front is about processing, preprogramming, and acting on that information: execution.

Like a city, the PFC has distinct precincts, horizontally and vertically organized; it receives from and sends messages deeply into regions of the brain's hinterland. Being the "final" associational cortex, it is disconnected from the primary sensory cortices' direct links to the outside world. It is our innermost sanctum. Firing almost continuously when we are awake, it's more quiescent when we sleep. The PFC differs slightly between the sexes, and widely among individuals—as much as fingerprints. Is the PFC different from the rest of the cortex? Edward G. Jones of the University of

California at Davis, the world's leading cortical neuroanatomist, says, "You know when you're looking at it, no matter where you're looking. And within the prefrontal cortex, you have these rather subtle, and sometimes not so subtle, differences, areas whose borders the practiced eye can usually detect. A novel thing about these borders, determined by the eyeball technique, is that when they are studied by other methods such as looking for connections, for example, these connections seem to stop right at the border where your eye detected an architectonic change. It's really quite remarkable."

Jones's comments arose during a conversation ranging the broad territories of his investigations. Researching this book, I have had encounters and conversations with men and women exploring the prefrontal cortex from many perspectives. Their voices inform this narrative. Sometimes I communicated with a person only once; more often, multiple times. There is, however, one irony. My inspiration for conceiving this project was Patricia Goldman-Rakic. Spurred by her compelling work of the late 1980s through the early days of the twenty-first century, I had a dialogue with Pat that spanned most of this time. So perhaps because I knew her work better than others', I put off interviewing her formally until later in the process. Then suddenly it was too late. She was gone.

For this reason, my neuroscience writing colleague Douglas Stein offered me the text of the in-depth interview he conducted with her at my behest as interview editor at *Omni* magazine. Although Douglas knew Pat's work as well or better than I and the conversation was vintage Goldman-Rakic, somehow the interview never ran. So we elected to share some of that conversation here. From both that Q&A session and my communiqués, the voice of Goldman-Rakic in chapter 1 comes from those intense days when she was approaching the peak of her powers and looking ahead to what the PFC might reveal about all of us.

At the end of the book is a full list of the dramatis personae and the dates of the most salient of (often) numerous comunications.

1

MEMORY
The DNA of Consciousness

With Miles Davis's *Kind of Blue* playing in the background, you focus on the computer screen. The rest of the world recedes as you juggle the numbers in your mind while your deadline looms. Suddenly, your landline rings, the front door slams, and your children race into the kitchen. The dog escapes; TV, video games, and stereo blare simultaneously. One of your printers jams. Your cell phone twitters. You search for some last-minute proofs, while opening an incoming e-mail. Do you have a strategy for what to do and in what order? Can you remember what you were thinking in time to act on it? Is your prefrontal cortex working properly?

Multitasking is a unique prefrontal talent that falls under the general rubric of "working memory." Working memory comprises the mind's intersynaptic DNA, its central operating system for thinking-in-time. Or to use another metaphor, working memory provides the musical notation system from which the higher brain's symphonies are composed. (Or to use another metaphor, it is something like cache memory in a computer.)[1] Yes, the PFC is the engine of choice, flexibility, decision-making, and foresight. But these functions are built on working memory's underlying action: holding information online. Working memory's dynamic processes drive PFC function across all time frames, at all levels of complexity, and in reasoned and emotional thinking. From the shortest-term memory fragment of remembering a telephone number to calculating advanced physics equations, masterminding a large corporation, or creating a large-scale work of art, music, or narrative, all partake of increasingly integrated levels of working memory to accomplish goals.

The brain bases of working memory have been discovered within the life span of a single generation of scientists, yet the origins of the concept are difficult to trace. The neuroscientist Karl Pribram may have been the first to wield the term in 1960. Or the phrase may have been adopted first by information theorists formulating computer programs for an artificial intelligence process that was an entrée into long-term memory, a short-term memory that is, as one neurobiologist crudely phrased it, "that fragile period when if you . . . hit an animal on the head with a hammer in the first twenty hours after it learns something, it won't remember it." The father of the term "working memory" is the British psychologist Alan Baddeley. In 1974, with Graham J. Hitch, he proposed a remarkably useful, if mechanistic, model wherein sentient behavior partakes of an "executive" function that controls behavior and two "slave" units that hold the relevant information "in mind" and available to the executive. Within this model, "working" was the operational word, stressing its dynamic nature, as opposed to the passive quality implied in "short-term" memory. But for decades most hard-core neuroscientists presumed the working-memory phenomenon was just like long-term memory, only shorter.

"Nobody in neuroscience knew about working memory until I started talking about it in the eighties," Patricia Goldman-Rakic declares in the early days of the twenty-first century. Goldman-Rakic, sixty-six, was killed in July 2003. Struck by an SUV on a street in a Connecticut suburb, she was jaywalking, probably thinking about her work, her PFC not attending to traffic concerns, having shuttled that processing to brain regions where automatic-pilot stuff is relegated. Her death was a stunning shock to the brain science community, and she was openly mourned in ways members of her tribe seldom are. Her campaign to bring the prefrontal cortex front and center where it could be explored with the full intensity of contemporary science has proven to be a triumph of passion, will, foresight, and determination.

Since the 1970s, Goldman-Rakic struggled to get working memory accepted as research worthy of establishment neuroscience. Virtually all wetware neuroscientists then viewed it as the conjurings of the "soft" psychologists. "Real" neurobiologists focused on long-term memory, the stuff stored in the attic of the mind. Long-term memory, whose essential activity, a kind of "stamping in information for archival purposes," as Goldman-Rakic once dismissed it, was then seen as antithetical to working memory. (Amazingly, even into the early 1990s, leaders in memory research omitted the prefrontal cortex from their anatomical diagrams—it simply wasn't wired in.)

Also irksome was that working memory was too complicated to study with their rather rigid systems. Many memory researchers, she claimed, were actually only studying stimulus-response conditioning—mere forms of Pavlovian training. "But human behavior," she declares to Douglas Stein during one long conversation in the 1990s, "cannot be explained by even an infinitely large set of conditioned responses." Or they confused learning by rote with long-term memory. Those who did study memory typically focused on the temporal lobe, never the prefrontal cortex. One player in particular—she snorted at the name of a well-known memory researcher—"called the medial temporal lobe memory system 'The Show,' the major leagues of memory research!" But long-term memory, she says, "could never ever explain this distinctive quality of intelligence: the ability to use the knowledge you've stored throughout the cortex to modulate your response to the moment." The big boys, she thought, either missed or denied the PFC's crucially human component.

"I was a bit surprised by the resistance at the beginning," she admits. By the 1980s, however, Goldman-Rakic sensed she was headed toward some remarkably unique neural system. "I saw that working memory, this elemental physiological function, is the equivalent atomic basis of all cognitive architecture. I felt we had the very essence of cognition!" Working memory is all about adaptability, she thought. Human behavior quintessentially involves new responses, changing constantly, based on information available at that moment. This ability to update information from moment to moment is what evolved with the frontal lobe and associative cortices in primates, and is further elaborated in humans: the capacity enabling us to base our choice of actions on experience and knowledge. Knowledge and experience are representations encoded in neurons. One must hold these *internal* representations online to guide behavior in the absence of *external* clues.

Goldman-Rakic wanted to know how individual neurons—the o's and x's of the brain's computing machine—communicated in the PFC. She set up experiments in what is called single-unit analysis, a painless electrophysiological procedure that allows a researcher to record activity simultaneously from numbers of individual neurons with ultrathin electrodes embedded in PFC areas. One records from neurons one by one to understand how populations of them converse with one another.

She and colleagues trained monkeys in various delayed-response tasks, sophisticated wired-up versions of Jacobsen's experiments of the early 1930s. Recording from ten, twelve, or twenty-four separate neurons in monkeys' prefrontal cortices, the team isolated groups of cells that fired during these specific delayed-response tasks. This neuronal activity indicated that the monkeys were storing information about spatial location after the cue disappeared and before they acted. The pattern of firing when the monkey used only a mental representation to guide its response showed Goldman-Rakic just how the neurons did the computation.

She was well aware that her test apparatus was a substitute for cues humans use to access information in working memory. "To you I could say, 'Remember the name of the restaurant you were just in'; 'the last five words I just said to you'; 'the last face you saw in the next room before you walked in here,'" she explains. "But I can't say that to an animal. Our presentation in the lab is a way of providing and controlling information, presenting it briefly, and seeing if the animal can hold on to it in a kind of scratchpad memory. Working memory is what you have in mind at the present moment."

Whatever task Goldman-Rakic set her monkeys to do, she saw that the delay response always worked the same. The common denominator was the inner image of the cue encoded in PFC neurons that remained active after the cue vanished. What once was present in the outside world now only existed in the inside world.

The mind's ability to create an internal representation of the withdrawn image, sound, or thought is central to this process. That representation in the mind's eye could originate either from long-term memory or from something you saw flash by the car window ten seconds ago. Some adaptations for auditory information may differ from visual or touch information, but holding the representation online in the absence of the cue is what's replicated across different prefrontal areas. That holding online of the representation is the basic DNA of mind.

Goldman-Rakic investigated the way PFC pyramidal neurons talked to each other: "Two parties are talking to this neuron, giving it information, and it may be building up its information. A single pyramidal cell can hold on to that information for ten, maybe twenty seconds. But an ensemble of neurons interconnected in a column can keep restimulating themselves, maintaining the conversation, and so may have some emergent quality keeping the information active longer. For a very fast system the time limit is much less. To hold on to the subject of a sentence while you go for the verb must be milliseconds," she speculates. "My constructing a sentence, or comprehending yours, requires a rapid integration."

A pyramidal cell in the PFC, she noted, works differently from one in other parts of the cortex. "A cell in the visual area, say, would just stop firing"—she snapped her fingers—"when the visual image disappears." Unlike the pyramidal cell elsewhere in the cortex, those in the PFC are not stimulus-bound. "That's the secret of cognition," she repeats. "How the cell holds that information? What is the nature of its input and output that give it that unique ability? These are the $64,000 questions."

Why are people's working memory abilities so variable? What are the prefrontal correlates of extraordinary talent? "Every person has delay cells in his prefrontal cortex," she says, "but some people may have wonderful cells that fire for fifteen seconds, during which time that person could integrate volumes of information. Take an arithmetic problem: you've got to hold all this in mind while you keep computing. Some eight-year-old will do all these mathematical gymnastics with incredible ease and accuracy. So if we were to put electrodes into a particular target, like the dorsolateral PFC, Brodmann area forty-six, we might find millions of cells that are exceptionally clean, sharply tuned, that have a capacity to hold on to information for, maybe, fifteen to twenty seconds. Whereas in another person those same cells will only do it for three seconds. "How working memory functions takes us in a direct line of reasoning to the underpinnings of intelligence. If there is a bell curve for intelligence, there is probably one for working memory capacity. There is a high correlation between performance on working-memory tasks and standard reasoning tests, and why not? Reasoning ability, simply put, is the ability to use representation to guide behavior."

Ironically, in light of the "mindless" accidental nature of her death, Goldman-Rakic once mused about how common, even comic, is the range of low-grade working-memory disorders we all experience. "Your frontal lobes fail you a lot," she reminisces merrily. "I often find myself making automatic responses that don't work. I started parking my car in a different lot, and had to take a different exit route from my office to get to it. Going to the left to get out of this building, which actually is a sort of radial arm maze, is the customary response I had made. Day after day for years I turned left at a choice point. Now I'm confronted by a new choice. Should I go left or right? Well, on three occasions I went left. I absolutely perseverated in my error! This is where the frontal lobe is so important: it overrides the automatic response. I didn't realize until far down that long corridor that I had to walk all the way back. My car was in another spot, but still I, the human with the PhD, was doing exactly what the frontally damaged animal does in the delayed-response task. He can't hold on to what he just saw. Lacking a representational system, he responds with the system that's not damaged."

There must be some neuronal dialogue, ensembles of cells conversing, each playing a role, like the keys in a chord on the piano. Jazz and improvisational piano playing obviously employ dynamic working memory. Playing extemporaneously, a pianist summons ideas from the well of the musical mind, but the self-generated ideas guiding these responses are not fixed notes, an inflexible score, but inner notations that change from moment to moment. The improvising musician never plays a piece the same way twice. Compare jazz to delivering a speech: "When I read the text of my speech," says Goldman-Rakic, "I've thrown a switch in the brain, turned off the PFC, and am using the sensory-guided mode of performance. Certain words are connected to certain vocal responses. But if I put down my text and speak extemporaneously, I'm constructing the speech as I go, and cannot even repeat the last sentence I made, because I have to erase it to move along. Working memory is like a mill; long-term memories are the grist for that mill."

The Perception-Action Cycle

Long before Goldman-Rakic began talking up working memory on the East Coast, Joaquin Fuster at UCLA had already published a paper demonstrating a unique firing pattern in the monkey dorsolateral PFC during a delayed-response task. Today a courtly, white-haired man, the Spanish-born Fuster has the quick, efficient moves of an athlete as he carries out research, teaches, and writes books. "I immediately thought the cells we saw were the mediators of working memory," he declares. Maybe the world was not yet ready for the discoveries Fuster and Garrett Alexander made in 1971. After all, Fuster published his findings years before Baddeley brought the first working-memory model to the table. In next three decades, Fuster elaborated his findings, presciently suggesting that this blazing activity in the PFC was part of a complex circuit involving many areas of the brain simultaneously, and several varieties of long- and short-term memory.[2]

Fuster had a unique insight into the special genius of PFC cells. In 1982, by training monkeys to switch between "what" (object) and "where" (location) memory tasks, he isolated neurons that selectively fire during each. That these cells are intermixed throughout the lateral PFC suggests that the PFC infrastructure integrates memory of object identity and location at the cellular level. He says, "In the prefrontal cortex, representations are highly idiosyncratic, very much related to one's experience—and therefore highly variable from one individual to another."

From the 1970s, Fuster increasingly realized that some prefrontal cells are multipurpose, their activity neither job-specific nor restricted to one sensory modality such as vision. These PFC neurons might instead fire to perform a variety of tasks, each calling for a mix and match of the senses—motion and object seen; sound and color; object and sound. Monkeys trained to associate a visual cue with an arm motion had neurons that fired only in the combined image-arm-movement context. Animals trained to associate a tone with a color showed prefrontal neurons that fired only to integrate both auditory and visual dimensions. Such cells, which treated the two streams as one twinned representation held in mind, are exclusively cross-modal. Audiovisually tuned for a specific set of sound and color, they perpetuate information about the "whatness" of this "soundcolor." No wonder we demand soundtracks with our video games, visuals with our MTV.

Fuster further saw that PFC neurons could form, dissolve, and reform their associations depending on the context. They were not committed to a discrete motor plan or sensory modality, or even the same polymodal associations. One could imagine these multiplex pyramidal cells as versatile freelance consultants—doing work for whoever called them up. Promiscuous even. "This was an entirely new concept when we first announced it. Activated ad hoc, yes!" Fuster exclaims. During that Stone Age stage of PFC research, Fuster alone understood this compelling characteristic of PFC neurons—that they can hang together to do a task, then disband or form other affiliations. The notion that prefrontal neurons are uniquely polymorphic powerfully influenced the next generation of PFC explorers.

The genius of the PFC neuron is even more impressive in its role as bridge over time. The PFC owns time. Working memory is essential for the "execution of successive acts in a structure of behavior over time," stated Fuster, seeing how PFC neurons fire during the space between the stimulus and the response, during the temporal gap when you memorize a telephone number and you punch it into the phone. To organize your actions, you need a neural mechanism to integrate them across time. If now this, then later that. If earlier than that, then now this: cross-temporal contingencies. This for Fuster was the unique prefrontal factor: he called it "temporal integration," marrying past and future across the gulf of now. By the early 1990s, Fuster had concluded that one set of PFC neurons are predictive, prospective cells that look to the future, while others are retrospective, looking to the past.

From this arises a third temporal dimension: the "memory of the future," as the Swedish neurobiologist David Ingvar named it 1985, the "I remembered that I plan to visit her tomorrow . . ." template.

A recollection of the plan to be executed can be as essential to one's sequence of actions as the straightforward preparation to act. Memory of the future is an extension of working memory at its existential bedrock. That is, no matter what the world's chaos, my own internal distractions or physical perturbations, I continue to remember what I intend to do tomorrow, next week, next year. This steadfastness of mind is embedded, for example, in the Latin verb conjugations of the future past participle, generally translated as, say, "I will have done [this thing] before sunset." Steadfastness of mind is something that characterizes the leadership of such heroes as Winston Churchill and Nelson Mandela.

Another prefrontal function is the role of attention over time. Obviously, there is an intimate relationship between working memory and attention. But what is it? Before scientists began picking it apart, "attention" was perceived tautologically: attention was attention. But to Fuster, attention comes in three flavors, quarks of attention. One is focus: the batter focuses on the ball as it leaves the pitcher's hand; in the airport you locate on the departure screen your flight number, gate, and time, and attend to that as you get a newspaper and coffee and go through security. Attention as keeping a representation of a sensory percept zeroed in over short time periods.

A second form is effortful attention. This is dedication, the drive that compels a person to persevere, keep striving, maintain discipline, and keep his eyes on the prize. It can be inextricably bound up with motivation, will, and desire. Attention with a capital A; attention over the long haul. The third attention is exclusionary, inhibitory. It repels the continuous sensory barrage to which the brain is exposed, and runs interference against distracting thoughts, and inappropriate behaviors and remarks. This attention overrides the habitual old groove that is such an effort to break out of—Goldman-Rakic's automatically walking to the old parking lot. Inhibitory control is absent in babies too undeveloped to curtail reflexive arm motions even when they want to; they lack motor inhibition. When brain damage to the orbitofrontal PFC causes the loss of this attention, primitive drives and emotions can gain the upper hand over reason and social conventions. As we will see, inhibitory control in cognition may be a prime indicator of IQ.

The prefrontal cortex's role in organizing action and the

"attentions" over time led Fuster to see that different categories of information in multiple brain areas play in working memory's big show. "Cells in many other areas of the cortex showed characteristics we'd also seen in PFC memory cells. We thought the entire cortex worked in concert, if you wish, for this form of active memory," he says. Thus Fuster was devising a theoretical model in which all memory, sensory impulses, and planning are interwoven in a giant, cyclical feedback engine he called the "perception-action cycle." Working, or "active," memory, as he calls it, is just one element in the big picture.[3] The PFC is the summit in the hierarchy of structures that form this perception-action cycle, integrating multiple inputs and outputs from many brain levels, translating them into actions that in turn produce changes in the environment, which are then perceived and analyzed in the posterior cortices and once again fed back into the PFC. Fuster's is an architecture of circularity: feedback and feedforward at every level in the ceaseless stream of reciprocal neural processing from spine to brow.

Surprisingly for a bench scientist, Fuster slides easily into the philosophical implications of his system. First of all, he says, it exorcizes the homunculus, the gnomelike puppeteer in the center of the brain. Instead of a miniature boss-operative, there are many semi-autonomous agents processing lots of information. "So it goes in a cycle in which there is no true origin, and therefore no need for a center for initiation of actions," he explains. "Because initiation of actions is a factor of the competition of small stimuli acting at the same time, many of which we are not conscious. So what you have is a statistical decision, a summation of impulses that we are not aware of, and to the extent that we are not aware of these stimuli, we feel free!"

So is free will, then, an artifact? Is self-determination merely the end result of summed computations, a calculus of neural events, or consensus voting among tiny unconscious impulses? "No, not an artifact," he replies, "but free will is a by-product of something which is to some degree deterministic." Since the work of the mind is unfolding in a statistical manner, Fuster thinks, stimuli soliciting an action are fiercely competing at any given moment. So the PFC acts as arbiter, awarding the stronger, winning impulse with conscious attention and intent to act. We may be aware of our intent to act, but not the vying neural competitors at work behind that awareness. We will see this idea of PFC "bias" expressed in the constructs of other investigators as they deploy computer models of the PFC to understand its special genius.

Gradients of memory, then, constitute a relational code emerging from the combinatory nets of neurons in the "grid" that fire and wire together to build a unique and dominant representation that biases and influences all statistical events over time. In this sense, Fuster says, one might consider one's "self" to be embodied in this one-of-a-kind web of neuronal relationships that fire together more frequently than other possible firing webs, and thus become the dominant web of neuronal relationships. The PFC's role is to manage the integration of such competing actions—outward movement, speech, and inner thought—over time. This is one hell of a model and it has inspired work on various levels of scrutiny. One scientist to pursue several of Fuster's observations is Earl Miller.

The Rules of the Game

In 2002, a reporter for German public radio separately interviewed Earl Miller and a professor at Harvard Law School for the same show. The radio producers then surreptitiously spliced both men's taped statements to create a phantom debate. So later, Miller was somewhat taken aback to hear in the midst of the fake face-to-face confrontation the law professor complain that scientists like Miller "think every mental state is attributable to a brain state!" as if this were a dangerously subversive idea.

"I was talking about executive control, and how information about rewards and rules encoded in the prefrontal cortex can lead to rational, goal-directed behavior," Miller reflects. "Then they interplayed this lawyer raving that it was all 'poppycock. Blah, blah, and that's what's wrong with neuroscientists is they think everything has a correlate in the brain, which leaves no room for free will, and if there's no free will, there can be no law! Because law is all about choice—choosing to be good or bad.'

"Well, of course mental events have correlates in the brain!" Miller blurts. "Unless you believe the mind is separate from biology somehow. And more to the point, neural correlates in the brain do not banish free will at all!" Given that there are so many ways of achieving a goal in this world, we need a brain that doesn't lock us up into one path of behavior. Free will is intact, precisely because every mental choice has a correlate in a brain that has the flexibility to confront these choices. "One has the responsibility to choose among them—this is what free will is all about. This is the essence of free will." Unwittingly, German radio chose a neuroscientist who in barely a decade has fit several significant pieces into the puzzle of how complex volitional behavior emerges from interactions between millions of neurons in the PFC. With a kind of relentless logic has Miller offered up one blockbuster experiment after another.

Now forty-three, as the Picower Professor of Neuroscience, Miller has a fistful of awards and his own laboratory at MIT. While his Web site photo shows him as a demonic figure backlit in a fiery red light, head shaved and do-ragged, face goateed, eyes burning embers, Miller is no iconoclast in his experimental techniques. He utilizes the same method of exploring the PFC as Goldman-Rakic and her mentor Walle Nauta: sending ultrathin electrodes into the lateral PFCs of monkeys to record firing from hundreds of individual neurons at once. But for this Miller takes a random approach. "We don't search for neurons that are engaged in the task, we just drop our electrodes down and record anything we find," he says.

Ear-stud bling and pirate beard aside, Miller is a precision thinker; the word "exactly" peppers his conversation. When he arrived at MIT in 1995, human imaging studies were beginning to take off, but he chose to stay with the old, uncool electrophysiology, because for examining the secret life of neurons, no imaging system was exacting enough. "Single neurons are the basic level of coding," Miller says. "I wanted to stay at that level because I'm interested in knowing exactly how information is processed and understanding the details of neural mechanisms that underlie executive control."

In 1997, Miller presented proof of Fuster's polymorph PFC cells. He and his team taught monkeys to mentally integrate the arbitrary relationships between objects pictured on a computer screen and their locations. When he recorded from almost two hundred neurons in the monkeys' lateral PFCs just after the images had vanished from the screen, he saw that many cells fired for a composite "what-and-where" construct in the monkeys' memory.[4] Such neurons are analogous to those in people that encode the memory of exactly what that golf ball nestled in that particular patch of rough looks like.

Miller also found that if the task only required the monkey to remember where the object was, the PFC neurons fired only for location. If the monkey needed only to recall the object's appearance, the same neurons fired for that image. The properties of many of the cells switched back and forth depending on the job requirement, suggesting, as Fuster proposed, that PFC neurons could change their tune depending on the score. About 50 percent of the neurons were cells that encode "what-and-where" relationships. "When object and location information are used together, as is typically the case in the real world," he says, information about these attributes converges in the PFC. "What-and-where" cells were the initial confirmation of Miller's growing belief that executive processes depend on the PFC's ability to fuse in one's mental universe uncommon relationships between disparate things.

"We are always figuring out relationships through experience and putting them together into a little model, sets of rules, logic or principles as needed to guide us through various situations," he says. How we do this "figuring out" depends on our capacity to forge from among wide varieties of information and mental representations the relationships that are new and arbitrary, relationships that evolution hasn't had time to program into our brains. Miller also saw that these neurons are distributed in both the dorsolateral and ventrolateral PFC (see figure 1 on page xi). "There may be gradients in the PFC where the ventrolateral is more 'what' and the dorsal part is more 'where.' But there is lots of overlap. And this is crucial, because it's the overlap that allows the prefrontal cortex to put together these arbitrary contingencies we need to learn new behavior."

Miller's discovery spurred him on to further challenge "temporary-storage unit" models of PFC function in which the brain's discrete sensory systems—vision, touch, hearing, and other parts of the posterior cortex—provide the PFC with raw material for short-term processing. This is the idea that the back brain "comes up with an answer," Miller puts it. "Then it's simply shoved up to the PFC and held online for a few seconds. We're showing that the PFC does something more, that it actually constructs the relationships needed by complex behavior."

To pursue this idea further, Miller's team then trained monkeys to remember pairs of associated images. An image of a house, say, would be paired with a picture of a flag. "If I tell you to remember a house and flag, and I say 'House,' you're supposed to remember to say 'Flag.' The monkey was doing the same thing but in the visual domain," Miller adds. Showing the monkey the house image, the investigators detected a rising activity in PFC cells that reflected not the house the monkey just saw but the flag image the monkey was anticipating seeing.

Clearly then, the PFC doesn't just receive inputs from back-brain visual systems and hold them online, but plays a command role in selectively extracting them from storage chambers and loading them in anticipation. Prefrontal cortex neurons generate prospective codes that allow us to prepare for events to come. "The PFC can play a role in anticipating things," says Miller, "and anticipation is what voluntary behavior and executive control is all about. You anticipate achieving some goal—preparing a fine meal or graduating from college—and yet you must be able to come up with the plans to achieve that goal." In the real world, sought-after goals are rarely achieved moments after we conceive of them. When we decide to go to the beach, we may realize we need our sunglasses. We have to recall (mostly unconsciously) what they look like and (more consciously) where we last put them. This ability to recall stored information in anticipation of its use, this prospective memory, Miller showed, involves PFC neurons that code for the "memory" of the anticipated, the expected but not yet occurring reality.

People pull up prospective codes for things that are not part of their actual remembered future: fantasies of winning the lottery or a Nobel Prize, acquiring a Lamborghini, conducting the Berlin Philharmonic, hitting a grand slam home run in Yankee Stadium. Different cues will elicit anticipation of delights in an unreal, alternate universe, constructed nonetheless into a powerfully detailed script from a wealth of hyperemotional imagery. With one caveat, Miller adds: if an activity is grabbing your attention now, the fantasy anticipation drama will not be running. Because the PFC is primarily an in-the-now processing unit, it is calibrated for present action or whatever is currently topmost in priority. But, he continues, "If you are not doing anything important, there's always gonna be this mode the PFC is in—anticipating things."

After isolating PFC neurons that hold arbitrary but convergent points of information, Miller's group next discovered neurons that encode rules. "Let me tell you a little about the back-and-forth cell," he says eagerly. Miller's lab taught a monkey simple rules using sets of pictures—analogous in human terms, say, to stop at red, go at green. The investigators picked new sets of pictures each day, so the monkey quickly had to relearn which picture meant go right, which meant go left. Not only did 40 percent of the lateral PFC neurons they recorded come to represent these contingencies, but a neuron only responded when the picture A meant, say, go right but not when A meant something else. Or only responded when B meant go left and not something else.

And it took the monkey only ten minutes to switch the rules. In that time, the neural activity in the PFC changed to reflect the rule changes. Here was evidence of the rapid-fire plasticity of a PFC rewiring itself to integrate relationships about a "seeing then doing" rule from information that is processed largely in separate systems in the posterior brain. And doing it with minimal training. The PFC neurons were showing off their agile, quick-break abilities to get a new rule into play.

People constantly learn arbitrary relationships, rules as elementary as stop at red, go at green. We are not born knowing the rules, but pick up protocols to play whatever "game" to maximal effect. Dining at a restaurant is one of Miller's favorite examples. You know the rules: how to access the menu, choose your drinks, order from servers, pay the check, and tip. While memories about dining in one particular restaurant on one particular evening are probably stored elsewhere, the PFC extracts the general features of previous restaurant visits and procedures to give you a general set of behaviors for eating out tonight. And it alters these rules so they can be customized for a bayou-side catfish joint or a four-star Chez Something-or-other.

This experiment showed PFC cells encoding concrete rules, where the rule is always tied to a specific stimulus—red means stop; green says go. Miller next sought the neural correlates of more abstract rules. Humans, and perhaps monkeys, engage in behaviors where the rules are more free-floating. A human calibrates his judgment and embarks on a course of action based on such concepts as "truth," "justice," or "fair play," even though they're not tied to a concrete agent. Would PFC neurons encode for these rules as well?

In this experiment, monkeys viewed two pictures, one after the other. If the "same" rule was in effect, the monkey indicated "same." Conversely, if the "different" rule was in effect, the monkey had to respond only if the two pictures were different. Miller's group trained the monkeys until they were adept enough to make judgments about "same" or "different" even if they were seeing the picture sets for the first time. Recording from neurons in the lateral PFC, they found that up to 50 percent of them conveyed information about the "same" or "different" rule. In fact, more neurons were concerned with the abstract rules of the game than with working memory. This suggests that rule-encoding and rule-representing is perhaps an even more fundamental PFC function than is working memory.

"The definition of an abstract rule," Miller declares, "is something that can be applied to a new experience for which there are no preexisting associations." The genius for fast, efficient, abstract rule-encoding frees an organism from getting stuck in the same old associations or rote behavior. It permits shortcut learning, enabling a smart animal to maximize his advantage in a new situation—think on one's feet—whether it is an engineer refitting the building codes of a site to the architect's revised plan, or a courtroom lawyer revising her examination style after a witness's sudden revelation during a trial. By their freelance nature, PFC neurons can encode for a virtually limitless numbers of rule-representations.

Continuing to explore the neural substrates of rules of the game, the lab looked at category-making. How do we fundamentally organize objects and experiences—apples versus oranges, raw versus cooked, liberal versus conservative, growth versus income stocks? Actually, what don't we categorize? How does the brain create category boundaries as the landscape of experience changes? Miller found that individual cells, "category neurons" as it were, in the monkeys' PFC become tuned to the concept of "cat" and other cells to the concept of "dog."

What grabbed everybody's attention was the design of the experiment. Miller's team collaborated with his MIT colleague Tomaso Poggio, whose lab created a computer-graphics 3-D morphing design program straight out of the *Terminator* and *Matrix* FX vocabularies. The experimenters took three prototype cats (a house cat, a cheetah, and a tiger) and three prototype dogs (a pointer, a St.

Bernard, and a German shepherd) and, digitally "melting" cat and dog characteristics together, generated animated composite images that were combinations of many possible feline-canine arrangements, from nearly pure cat to nearly pure dog. By blending differing concentrations of cat and dog in series of images, they could vary the "catness" or "dogness" of an image and push the limits on category boundaries.[5]

Watching the image on the screen morph from a cheetah to some indefinable entity, then resolve into a St. Bernard, it was hard to pinpoint exactly when the creature was no longer a "cat" and now a "dog." I was worse at it than the two lab monkeys, but then again I hadn't trained like they did. Working for months, the monkeys, who had never seen a live cat or dog, learned that any image that was more than 50 percent dog was dog; any image more than 50 percent cat was cat. The monkeys had become skilled enough to tell when an image was 60 percent cat and 40 percent dog. Since the program generated many new cat-dog chimeras during the experiment, the animals weren't just rote-memorizing specific image mixes.

Beforehand, Miller wasn't sure what he'd see going on among the PFC neurons. "We knew categories had to be represented somewhere in the brain, because monkeys use category information to guide their behavior. But I thought it was possible, even likely, that we would not find evidence for category representation at the single-neuron level," he admits. "I thought it was likely that to represent a 'cat' category, there might be neurons to encode whiskers, ears, tails, neurons for overall shape. I suspected that somehow, at some high level, all these neurons might respond at the same time to amount up to the category, cat."

That's not what they found. Recording from around four hundred cells, they observed nearly one-third to be specifically category-responsive, those firing to all-cat images until the image morphed up to the edge of the cat-dog boundary; others firing to all-dog images until the image approached the dog-cat boundary. Once the image crossed the species boundary, firing activity changed abruptly. "It was sharp," describes Miller. "One window opening, another closing—just like that."

Cat-category neurons responded to every manifestation of cats. So two cats could look very different from one another and the

PFC cells still treated them as "cat." Or one dog might resemble a Doberman and another a dachshund and the PFC would say they're both dogs. "In the end that makes sense to me," Miller offers. "Because when you walk into a room, you instantly recognize a table, a chair. With enough experience, a category gets encoded on the single-neuron level, allowing you very rapidly, efficiently, and effortlessly to organize and conceptualize the things around you."

When we perceive things noncategorically, objects or events can change gradually, shade or evolve smoothly from one to another. The sharp boundary effect, the sudden switch-off between dog neurons and cat neurons, however, fits our experience of category-making. "So, as with that sharp behavioral boundary where in our minds we know something is either/or, we expected to see some sort of sharp boundary in neural activity. And that is exactly what we found," he continues. Pondering the "street" implications, one might see the beginnings of an explanation for why political and ethnic problems are so intractable. People easily form sharp, arbitrary category boundaries between "us" and "them," categories that are fed by emotional wellsprings to the extent they are hard to unwire. It may be more difficult to break down a category boundary than to build it.

But do category-forming neurons exist solely in the PFC? Another place to check for these cells is the inferior temporal cortex (ITC), that region just above and behind the ear involved in the high-level processing of visual recognition and visual memory (see figure 2 on page x). If you take any neuroscientist off the street who's familiar with memory and ask him where categories are going to reside, Miller says, he'll tell you it'll most likely be the inferior temporal cortex. When Miller and his crew "marched back" in the cortex, comparing neuron firing in the PFC and the ITC while a monkey played the cat and dog game, they saw some neurons that conveyed implicit information about the category, as in the PFC. But just as often, ITC neurons conveyed information about cats and dogs as individual animals. If the monkey viewed two very different-looking cats, ITC neurons might convey information about them being in the category of cat, but also might prefer to fire for certain cats over others.

But that's not all. Yes, long-term memories for abstract categories of, say, cat families may be stored in the temporal lobe, but they are mixed in with all sorts of other information about fur, whiskers, paws, a grab bag of physical attributes of what individual cats look and act like. Prefrontal cortex neurons, on the other hand, convey the categorical equivalence and ignore differences of appearance. "That's a definition of a category: I can treat a tiger and a house cat as both cats even though they look different from one another. We only see that equivalence across changes in physical appearance in the PFC. I didn't expect to see such striking differences in the two areas," he exclaims. "And that the final level of abstraction only takes place at the level of the PFC."

A PFC/ITC category-generating network makes sense operationally. Information about the physical appearance of things is fairly immutable, fairly hardwired in the temporal cortex. Individuals always look pretty much the same, with a little variation over time. But categories are more ephemeral. "I could have a transportation category," muses Miller, "and can instantly generalize upon and modify it to include a new form of transportation, such as the Segway scooter, that two-wheeled 'human transporter.' High-level abstract categories and concepts need to be more dynamic and fluid."

If categories were stored back in the temporal cortex, along with information about the physical appearance of the Segway scooter, one would need information about its motor, wheelbase, steering mechanism, and battery connections stored alongside the details of every possible kind of transportation. By waiting until the last possible stage of processing to encode the abstract category, you can be much more open-ended with what you regard as a member of that category. It's a brain being efficient after millions of years of evolution.

Another question needed to be answered: that, having evolved on a savannah where dog and catlike predators roamed, did the monkeys have some "genetic memory" of cat and dog? To test the genetic memory hypothesis, they reassigned the cat and dog stimuli to three new arbitrary categories that had nothing to do with cat versus dog. "The monkeys learned these arbitrary categories just as easily as the original cats and dogs," says Miller. "Further, all the PFC shifted to reflect the new arbitrary categories. This is a strong hint that even the original categories were learned and arbitrary to the monkeys."

Miller replaced cats and dogs with numbers of dots—abstract entities. The number experiment, dubbed rather inaccurately by the media "Monkey See, Monkey Count," involved another Herculean training and design effort that included preventing the monkey from cheating. The animal might simply memorize all the possible combinations and patterns of images. With a hundred versions of each number-dot picture—five hundred new stimuli every day—the monkey couldn't possibly memorize them all.

Judging the relative quantity of low numbers is highly adaptive. Many animals do it. It's a way of quickly categorizing and making sense of quantity. The task was limited to no more than five numbers, because after five or six items the monkey's ability to categorize number quantity drops off. In fact, without the verbal encoding power of counting 1, 2, 3, 4 . . . 100 . . . 234, and so on, a human's ability to conceptualize number quantity is not much better. If humans are prevented from verbally counting, they show the same drop-off in ability to measure quantity at about 5 or 6. And despite the headlines, monkeys can't count.[6]

Recording from lateral PFC neurons, the investigators again found that about a third were tuned to a specific number. Firing intensity, furthermore, progressively declined as the number of dots moved away from the neuron's "favorite number." Overall, the neural patterns seemed to form a "bank of overlapping filters." The neurons "knew" that the quantity of three is closer to four than it is to one. "The results from our cat and dog and number studies are remarkably parallel. The final representation of the abstract concept seems only to come to fruition in the PFC," says Miller. But there were subtle differences between number and cat/dog processing.

"We did the number experiment because numbers, although very abstract, are an example of a genetic memory," says Miller. "Many, maybe most, animals can make small number judgments without explicit training. What we found recently is that the innate memories for small numbers seems to be stored a small area in the parietal cortex and is then 'loaded' into the PFC when needed. This is in contrast to the cat and dog categories, which did not seem to be explicitly represented in sensory cortex and underscores the importance of the PFC in learning arbitrary rules."

Your Inner Proust

How the PFC distributes working-memory computations across its various territories is the subject of much research. The ventrolateral PFC is "like day and night to the dorsolateral," asserts Michael Petrides, the director of the Cognitive Neuroscience Unit of the Montreal Neurological Institute at McGill University. Petrides, among others, proposes the existence of "multiple executive processing modules" within the human PFC. With the arrival of functional magnetic resonance imaging (fMRI) techniques in the mid-1990s, he began digitally capturing the contributions of these different subsectors, and devised a model of the lateral PFC as a two-stage mental processor with mastery over the flow of experience during time.

For him, there is a "looming dichotomy" between what the "upper" and "lower" lateral prefrontal areas do. "I'm claiming there is a big chasm—as big a divide as between North America and South America—where dorsolateral areas do something predominately different from the ventrolateral ones." In Petrides's model, the ventrolateral PFC, Brodmann areas 47/12 and 45, constitute a kind of search engine for retrieving specific memories and data from archives in the posterior brain, particularly when the information is embedded in an ambiguous context or interleaved with many other memories. When we need to recall something in particular—Who was that man at the party?—the mid-ventrolateral PFC is recruited to the search.

A person with damage to the ventrolateral PFC has lost this capacity to initiate an archival search for that one piece of information—where the mind serves as a heat-seeking missile targeting the exact name, number, image, or idea. "If you are a patient with frontal damage," Petrides says, "you are not amnesic. You're also just as smart as anybody else, but you make mistakes; you fail to retrieve information, not because the information is not there, but because you lack the basic executive processes that enable you to go back into your memory traces when the retrieval is tricky and highly ambiguous."

This active, precision-targeted recall is distinct from varieties of remembering that do not require the ventrolateral PFC's talents. Petrides continues, "If I meet you for the first time, then later run into you, I will recognize you. My posterior visual, auditory, and multisensory areas will process this information, as long as they have good interaction with my hippocampus and other limbic areas. So if I see you again, similar images will be reactivated in posterior cortical areas, and they will be sufficient for me to recognize you. For this I don't need my frontal cortex." In people suffering from damage to the ventrolateral PFC, this passive recognition memory system is usually intact.

This nonprefrontal, associational memory system can be activated by a single, momentary concordance. "I could be having a nice martini somewhere in downtown Montreal," Petrides goes on, "and I start thinking about a particularly strong experience of a meeting in Colorado where we had a great fish dinner. I then immediately remember that you were there. Or I might have been watching TV and there was something about Denver in the news. That immediately takes me back to the meeting, and I remember the restaurant where we had the fine seafood dinner and what great fun it was, and suddenly the whole image springs out in my mind of you and others sitting around the table. That is what happens most of the time. Our memories are being reenacted, retrieved because one sees things again, and so new traces touch old traces. One association triggers the second and so on.

"And yet, as soon as someone asks, 'Was she at the Colorado meeting?' in this active kind of memory, I can't merely let memories link to other memories. I initiate a search of my memory traces for the specific pieces of information I want." The strength of an active initiatory memory system subserved by the ventrolateral PFC and its networks has implications for the contours of the individual self. Take Marcel Proust in *À la recherche du temps perdu*. Obviously Proust, intensely attuned to his associational memories, also had an extraordinary search engine for his archives of lost time. Proust's "faithful guardians of the past" could roam his back-brain libraries for "texts" of amazing specificity, and retrieve them in consummate detail. Such gifts are invaluable to any art or calling. Take the physician who, with scant evidence of symptoms, can search for and summon up the probable diagnosis of a exotic skin disease based on data about it stored in her long-term memory.

Your Inner Palm Pilot

On the other side of Petrides's lateral prefrontal divide is the mid-dorsolateral PFC, Brodmann areas 46 and 9. As the PFC's executive suite, the dorsolateral is a specialized place where memories and events, once stored and interpreted in the posterior cortical areas, are now summoned to be sorted, monitored, and recalibrated. The mid-dorsolateral PFC's job is to attend to and manipulate the status of mental events on our assembly line—actions we intend to take, plans we expect to execute. It is home base for our "memory of the future."

"At any moment in our lives many things compete for our current awareness," Petrides explains. "A person might be holding online a number of relevant events and cannot afford to ignore any of them, but must monitor them all. He must keep track of which events have happened, which events are yet to happen." This manipulative ability gives us tremendous flexibility. As the beginning of strategic planning, it is essential for making creative designs. "I wake up and set up six or seven intentions," says Petrides: "'This morning I have to call Mrs. X, or be in my office at three o'clock, when she will call. I also have to make sure I contact those four others about the party today.' In the mid-dorsolateral area, neurons have coded these intentions. So as I go through the day and do many different things, those neurons continue to code them."

After he calls Mrs. X and checks off that as "done," his mid-dorsolateral PFC recodes the neurons. The mid-dorsolateral PFC thus reorders a series of events being held online. "How can I do any manipulation, if I cannot hold those relevant six things in my mind, if I cannot say A, B, C, and D are relevant components?" Petrides asks. "And that A has moved down to B's place in the list, and now C is in A's position? I must have a mind capable of attending simultaneously, keeping track, and prioritizing multiple representations."

People with damaged mid-dorsolateral PFC areas fail at keeping track—whether of objects in a sequence, numbers, or abstract ideas. These patients do not lose their capacity to speak or remember, but inevitably their lives collapse. "They are intelligent on educational tests, but not street smart," says Petrides. "To be street smart means that right now you are attending to one thing, while at the same time there are little lightbulbs keeping track of the three other things that also need to happen. So at the appropriate time you can quickly turn off A and turn your attention to B and C. Patients with mid-dorsolateral PFC injury cannot do that kind of prioritizing or self-ordering." They cannot easily get out of groove A. This dorsolateral ability no doubt evolutionarily preceded techniques humans have invented to enhance this executive function: from the invention of writing itself, to its simple subrule, taught in grade school, to "outline" our papers. It is the mental construct behind, perhaps, everything from the Dewey decimal system to esoteric electronic stratagems for organizing vast amounts of material along space and time continuums.

Area 46 may be an organizational sector. But is this dorsal/ventral dichotomy as clear-cut as Petrides has postulated? The Cambridge neuroscientist Adrian Owen, a sometime Petrides collaborator, agrees that the mid-ventrolateral PFC retrieves specific memories. "If I ask you to remember the number 7946, the ventrolateral is involved in retrieving that number later on," he says. "But if I ask you whether there were any *even* numbers in that sequence, then you must introspect, work though the contents of your memory and decide, yes, there were even numbers: four and six." This, Owen thinks, requires the participation of the mid-dorsolateral PFC.

In imaging studies of the frontal lobe, Owen has tried to "really crack open this ventrolateral/dorsolateral thing." He now thinks a "gradient" exists in levels of processing complexity, with a gradual change from more basic memory processing at the ventrolateral level to higher-level processing at the dorsolateral stage. "The dorsolateral PFC," he suspects, "is involved in identifying potential strategies to facilitate, or make memory most efficient. The dorsal/ventral distinction, then, would be for me now, one of levels of abstraction." So then is the dorsolateral PFC involved in computing more intentional, and therefore conscious, processes? To Owen conscious versus unconscious may not be real distinctions. "When somebody is looking at the contents of their memory, are they aware they are doing it? Or with higher-level thinking, when you figure out a way to approach a problem and then set about doing it, how aware are you of your scheming? Who says to himself, 'Well, now I'm going to start strategizing . . . '?"

In a test of shape memory, Owen used abstract designs rather than familiar shapes, to avoid a situation where people would say, "I remembered the square." This situation is not unlike Miller's need to keep the monkeys from "cheating" at number tests. Yet the subjects reported afterward that they remembered the shapes nonetheless, by creating strategies such as "more shapely" or "less shapely." They were not aware of creating schemes to facilitate recalling these abstract images. On the scanners, Owen saw elevated dorsolateral PFC activity. So, he argues, this region may serve "to identify order in the world. It says, 'Yes, I can use shapeliness to facilitate memory!' This strategizing is not something we necessarily do in a conscious, self-motivated way, but it is the way this brain region is set up to maximize effectiveness."

Owen also has "some really nice data" to foil Mr. Homunculus. "This model avoids that little brain-within-a-brain problem in the sense that the dorsolateral PFC relatively automatically identifies high-level structures in the information it is processing." Owen suspects this automatic-ordering faculty is almost always based on past experience. A square is always a square with certain geometric properties, but that's not true of all objects we see and think of as shapes. We may try to organize new shapes to fit into categories we've seen in the past or are familiar with. Owen's idea—that there is an innate bias toward ordering the novel and random flow of the external world—concurs with Miller's category-building neurons. Just how we compose these organizational strategies is highly idiosyncratic, intensely personal.

Take a chess player doing a spatial memory test: this person will tend to refer to objects in space in terms of chess positions. "The chess player is well practiced at spatial thinking. But I'm not a chess player," says Owen, who is in fact a lead singer and bass player in the rock band YouJumpFirst. "I don't see chess positions anywhere in the world! Now, if both of us are looking at the same spatial problem, our visual cortices will do exactly the same thing. But the 'strategy' we'd bring to bear on that problem would be entirely different. Because people do organize reality differently, it's difficult to find specificity in the prefrontal cortex. We talk about PFC function in terms of 'manipulation' and 'monitoring' strategies, but what are these? We all in the field think we know what we're talking about, but none of us actually believes that this is solely what this region does."

But when does the PFC "know" that it needs dorsolateral intervention in order to conduct a more complex manipulation? And is there a limit to the strategic operations it can keep online at any given time? The Stanford psychologist John Gabrieli, using fMRI to explore working memory, was struck by how even modest tasks, such as remembering a string of letters or digits for a short time, engaged huge portions of the PFC. He was also impressed by the limitations of working-memory capacity—that it can only hold around seven bits of information at once. This basic unit capacity is actually less—more like three or four bits. But you can manipulate a couple of these units to add up to the "magic seven." "You're kind of saying to yourself: 'those 3' and then 'these 4.' You are juggling two things, and that suggests it doesn't have to be that hard a task before a lot of the dorsolateral PFC is involved," he says. Perhaps that's why phone numbers are generally seven digits, plus an area code.

"You begin to wonder what's going on in more complex executive operations. When does a quantitative thing become a qualitative thing? Once you get past about three items," Gabrieli adds, "it's as if your brain says, 'Okay, now I need to turn on this other computer.' That is true of almost everything in life. If you carry one or two shopping bags, it may seem as if the third is just another one. But that's when you start dropping things. What difference will one more make? It's a qualitative increase; at one point it becomes the final straw. If you have to manage enough information at once, it's simply that managing it becomes a dorsolateral executive process."

The Brain's Conflict Monitor

A growing consensus thus implicates the dorsolateral PFC as involved in identifying potential strategies to facilitate working memory. But "who" alerts the PFC to summon its special talents? To explore this key issue we need to visit another subsector of the PFC, the anterior cingulate cortex. Before that, however, we need to acquaint ourselves with the Stroop test.

It's worthwhile to stop and admire the Stroop test. Cited and applied thousands of times during the past seventy years, it is the classic examination of attention and lack thereof in the prefrontal cortex. No one can talk about working memory and executive control without sooner or later encountering the Stroop. It works this way: a player is presented with words for colors (e.g., GREEN) printed in either the color the word indicates, or another color (e.g., RED). On command, the participant must either name the word, or disregard the word's meaning and name the word's color. What's so remarkable is that when the print color differs from the word's meaning—if, say, the word GREEN is printed in red ink—a person takes longer to say "red" than he does to say "green." To name the word GREEN as "red," the person fights to inhibit and suppress the stronger tendency to say "green." Or he makes an outright mistake and says "green." "Scratching the itch," as the neuroscientist Jonathan Cohen put it. The error is called "Stroop interference," or simply "the effect." Others have likened trying to name the word's color to wading knee-deep in mental sludge.[7]

We are programmed to read words for their meaning. Thus when asked to suppress this response in order to focus on a word's color, our minds balk at this violation of what we "always do." Thus the Stroop neatly demonstrates a core function of executive control: the ability to override a strong but wrong signal to select a weaker but right one. Patients with PFC impairments, including attention deficit problems, schizophrenia, and various injuries, struggle with the Stroop. The Stroop is sensitive to subtle changes in normal brains as well. Fatigue, loss of sleep, minor brain damage, and strange environments, such as high altitudes, increase one's error rate and the time it takes to name a word's color. To test mental flexibility, the Stroop has been given to people in all sorts of extreme states, including climbers nearing the 8,000-meter mark on Mt. Everest.

The Stroop has escaped the lab in other ways as well. Recently it was programmed into the MiniCog, a little handheld electronic device, used by NASA astronauts. Its developers claim that corporate strivers, as well as space walkers, can check on their prefrontal CEO abilities at any anxious moment by seeing how they score on the Stroop. A Web site advises stock market day traders to practice the Stroop. Since they face an "oppressive opponent within their own minds," the ad warns, they can better cope with the constant bombardment of distracting external stimuli by practicing the Stroop. You will learn to better "filter what your brain deems unimportant, based on criteria you have given it." John Ridley Stroop published his invention and its first test results in the *Journal of Experimental Psychology* in 1935, the same year Carlyle Jacobsen announced the results of his chimp studies. Compared to Jacobsen, Stroop's name and experiment is far better known.[8]

In 1986, Jonathan Cohen, now director of the Center for the Study of Brain, Mind, and Behavior (as well as codirector of the new Institute in Neuroscience) at Princeton, was one of an exotic breed of young researchers captivated by the potential for applying connectionist computer modeling to the neurobiology of thought. At Carnegie Mellon he studied with the neural-net pioneer Jay McClelland. In McClelland's class, Cohen met Kevin Dunbar, whose previous work focused on the Stroop. Cohen vividly recalls sitting in McClelland's office when, Dunbar said, "If this connectionist stuff is so good, we should be able to model this Stroop finding."

"Neither of us had much experience in modeling," recalls Cohen, "but I wanted to try to build a model of the Stroop." In the mid-1980s McClelland and a few others were exploring parallel distributed processing architecture to simulate brain activity. They called these programs "connectionist" because, like actual neurons, their computerized simulated cells communicated with other simplified digital "neurons" in the model to create networks that in turn simulated brainlike behavior.

Over the course of 1987, Cohen and Dunbar went about designing a connectionist model of a neural network that could negotiate the Stroop test. They programmed in two processing pathways: one devoted to word, the other to color information. Both pathways would converge upon command to respond to a task demand: name the word or name the color. Like a human, the model had to select between the two competing processes—word or color. To mimic the human condition, the scientists strengthened the model's word pathway by "training" it more intensively than the color pathway. When they finally ran Stroop simulations, sure enough, the machine performed faster in "naming" the word than it did the word's color. Since in computer modeling everything is modifiable, they reset the program, overtraining the color pathway. Then when they ran the Stroop sim, the machine did better "naming" the color than it did the word.

"Out of this simple neural-net model leaped not only the fact that the relevant strength of the pathways could determine the speed but that everything was subject to control. Realizing the model could account for the basic Stroop effect, I simulated new learning data, and showed it could account for other findings. It was not perfect. The model had idiosyncrasies and unexplained elements we were not comfortable with. But," Cohen adds, "it provided a conceptual grounding for me."

Around that time, Cohen attended a conference on the burgeoning ideas about prefrontal functions. People were talking about Goldman-Rakic's work and recent findings about working memory. There was palpable excitement about the notion of maintaining information online. Inhibition—the PFC's ability to curtail rote in favor of new behavior—was another theory. "With all this percolating in my mind," Cohen recalls, "I started thinking, 'Maybe the prefrontal cortex is involved in the Stroop effect.'"

In doing the Stroop correctly, you are maintaining in your mind the rule, representation, or strategy. Your brain chooses the desired but weaker interpretation, inhibiting the stronger but undesirable interpretation. You want color, not word. But still, it's ambiguous. "I suddenly realized that the PFC might be sitting there presenting this information, not just holding it online, but using it to literally guide how the rest of the system will perform. And that epiphany was basically ten more years of research!" Cohan admits. He suspected that the PFC was weighing competing representations and judging which among them to give the go-ahead signal. It was not unlike Fuster's idea of competing systems adjudicated by the PFC. But how did the PFC "know" about this conflict in the first place, in order to attend to it and steer the neurons toward the right "goal"? That question led Cohen to a "lower" part of the prefrontal system, the anterior cingulate cortex (ACC).

The ACC (see figure 1 on page ix) is the elephant to the neuroscientists' blind men; everybody's got a slightly differently take on it. The mental operations it putatively engages in include heightening skin, touch, and pain sensitivities—even emotional pain. Thought to be a regulator of positive mood, it has been dubbed the brain's "cheerleader." Some think the ACC functions as the brain's quality controller or "oops monitor." Studies find it more active when a person lies than when he or she tells the truth. When dysfunctional, it may play a role in depression, obsessive-compulsive disorder, anorexia, and attention deficit disorder.

Because of its position and extensive hookups, the ACC, Brodmann areas 32 and 24, has many strategic alliances with other city-states of the brain. It is intimately connected to limbic structures and the autonomic nervous system that oversees heart rate, blood pressure, metabolism, and other of the body's housekeeping functions. "The part of the ACC that lies in the rostral prefrontal cortex," says Cohen, "is clearly involved in higher-level executive functioning. Exactly what that functioning is, is what's really interesting. "In my thumbnail account, this strip of cortex's mandate is taking stock of the system's performance. Internal states are the focus," he stresses, totally oblivious to the jackhammers pounding away in the Princeton psych building's throes of renovation. "The ACC is about looking inward, into whether the thing you are doing now, or about to do, will lead to a good or bad outcome. This information could pertain to motor performance or autonomic inputs—your stomach growling tells you your behavior isn't satisfying a fuel need. It may pertain to how you perceive you are doing on an SAT test or job interview."

More important, the ACC, in Cohen's view, is a chief player in control processes, those neural mechanisms that help the PFC adjust to changing demands; to reconfigure the amount of attention needed to think something through efficiently. That the brain somehow detects and monitors its own inner performance was until recently not a direct object of inquiry. The control aspect was assumed. Somehow the brain "just knew" when to turn up or down its intensity level. But this explanation was dismayingly "homuncular."

Since medical school, Cohen had wondered: what happens when a person decides to attend to "this" as opposed to "that"? At first he suspected that monitoring was a lateral prefrontal job. But he and his collaborators decided to see if other brain areas signaled to the lateral PFC what it should be paying attention to, or how much attention it should allocate. With Cameron Carter, then at the University of Pittsburgh, Cohen developed a computer model of an ACC to strap onto his Stroop-playing virtual PFC machine. The idea was arduous in its development. A breakthrough came when the two men recalled a brain-mapping meeting where the host opened the session with a rhetorical question: looking at the thousands of imaging abstracts submitted, which area of the brain was most active? It turned out to be the ACC. Indeed, the ACC fired up across many different studies—in response solely to the task's difficulty. The data were hinting that when the going gets rough, the ACC gets going.

Now armed with evidence that the ACC responded in some ill-defined sense to "brain sweat," Cohen and Carter were struck by reports that it might serve to monitor errors. As a task grows more difficult, a person makes more errors. But there was a knotty problem with the error theory. Much of the imaging data found the ACC to be active when subjects performed difficult tasks but made no errors, or no more than when the tasks were easy. The discordance between the error story and ACC activity piqued Cohen's interest. "Brain imaging has often been dissed for telling us things we knew," he says, "and here's a great example of it not only telling us something new about the ACC but providing evidence that forced us to think hard."

Subjects who performed flawlessly on the Stroop test still showed elevated ACC activity. Cohen asked himself, "What's going on in the Stroop that monitors difficulty?" It dawned on him: "Maybe it's not error but uncertainty." It made sense that a region surveilling errors should be most active when the task is most difficult and uncertain. "Difficulty and uncertainty are in part indexed by accuracy, at least your performance is," Cohen says. "Difficulty and uncertainty are what we ultimately came to articulate as conflict. Conflict is what's driving the ACC."

An attractive feature of the conflict hypothesis was that it might exorcize the homunculus once and for all. Conflict, not "My bad!" would be all the brain can know. There would be no "wrong" buzzer squawking, but a kind of neuronal dissonance, when the brain struggles to choose between two or more responses when it can only make one. Cohen was willing to go out on a limb and say, "When something is incompatible in the brain, there arises from the ACC a high activity that flashes 'conflict!'" Useful for gauging when the PFC needs to come online and when it doesn't, the ACC, then, could be the region that alerts the PFC to be alert for "incoming." Or to stand down.

So Cohen went back to plug an "ACC conflict monitoring unit" into his Stroop machine. With then postdoc Todd Braver, he set up the electronic units so that for each Stroop trial the "ACC" would gauge input from the rest of the nets and compute the amount of prevailing conflict among the "response units." Running the model, the two found that the computer's "ACC" did indeed detect conflict during the color-naming condition of the Stroop test, even when the model performed correctly. Tweaking the difficulty of the computer trials, it became apparent that whenever the model took longer to respond, it was because competition persisted between the alternative responses. This confirmed their hunch that the ACC was monitoring conflict, crosstalk that arose in the model's "word-" and "color-naming" pathways before the model made its response.

A basic tenet of information processing, hammered out by the designers of parallel-processing computers decades earlier, is that a computer program needs excess control in situations where there is crosstalk. Most artificial intelligence (AI) programs have control-alerting functions. If reduction of crosstalk is a primary function of control in parallel computing systems, then a brain, too, might monitor for the presence of crosstalk to know where it needed to allocate control. Such a monitoring signal from the ACC monitor would note either the presence of conflict or that the coast was clear, and quiet down until it detected conflict again, whereupon it would again recruit various degrees of PFC involvement. "The notion of a loop using conflict monitoring to determine how active the PFC should or shouldn't be," says Cohen, "began to make sense."

In feedback-feedforward loops, frequency and time are critical elements. When a person does the Stroop, says Cohen, "you are sitting there, primed before the stimulus comes in, ready to trip the response and quickly say the word's color. The (red) word GREEN appears on the screen coupled with a little bit of noise in the system, which could mean that you got distracted by someone slamming a door down the hall or something. And you utter 'Green!' And you go: 'Oh, that's not what I meant!' And you may correct it on the next trial."

Cohen designed the computer model's response units to "hover" at the threshold, barely below the levels that noise wouldn't trip them over. When the model made a mistake, it was because the stronger support for the word GREEN caused the green unit to respond before an "attention unit" had a chance to kick in to suppress it in favor of the color RED unit. The computer went for GREEN. And the ACC/conflict unit signal appeared.

Over a series of trials, RED response units begin to accumulate "strength," eventually overtaking GREEN response units to trip the correct response. But during an intermediate period, where GREEN still had a minor advantage but RED won out, there was a warning of impending conflict. Did human EEG studies of the Stroop, Cohen wondered, show signs of a brain wave that fired before someone did a correct task, as in his machine? Lo and behold, in the literature was an electrophysiological response called the M2C, a firing pattern evident about 200 milliseconds prior to the stimulus. "Exactly where we predicted is a conflict signal." (Interestingly, one-fifth of a second is about the time it takes for a batter to resolve his conflict about whether he's looking at a fastball or breaking ball and make the appropriate choice to swing or not swing.)

To Cohen, this M2C and the error signal were identical. "We say both reflect the detection of conflict. Conflict precedes response, and if you answer correctly, by definition it gets resolved in favor of the correct response. You've suppressed the incorrect response, end of story. On the other hand, if conflict results in an error, you continue to process information, which in turn leads you to acquire information about the correct response that competes with the previously activated incorrect one. You see the conflict and you correct yourself."

Cohen and colleagues next attempted to simulate how the brain fine-tunes performance to lessen errors and maximize "winnings" over time. They used as a template a long-standing finding called the "Rabbitt effect," so named for the British psychologist Patrick M. A. Rabbitt. The Rabbitt effect is a fairly commonsensical feedback loop: after making an error, you tend to slow down, be more cautious, and so become more accurate on subsequent trials. Then the better you do, the faster you go, and consequently the more mistakes you make, so you slow down. And so on. (Statistics on motorists' frequency of getting speeding tickets may confirm the Rabbitt effect.)

Cohen's critics have noted that the Rabbitt effect was meant only to describe explicit, conscious errors. Certainly, some explicit knowledge of having made an error has nothing to do with conflict—just "I didn't get it right," which leads you to adjust your performance. Cohen, however, contends that the ACC's conflict monitoring yields the more accurate results over time. And it may do so implicitly, unconsciously. Also, in "Rabbitt fashion," people do better when they have a string of hard tasks than if they have a string of easy ones followed by a hard task. Again the reasoning is commonsense: If the previous task was difficult, you will be more focused and conservative in your approach to the next one. If the previous series have been easy, you are lulled into complacency, slacken your focus, and so make an error.

The question for the Stroop computer was, again, one of feedback. Could the conflict signal alert the PFC sim to be more "on its toes" in the subsequent trials after an error? And slacken off after it seemed the trials were growing too easy? Cohen's group ran trial after trial to determine how much the model had to turn up or down the "alertness volume" to maximize its correct responses. They found that the control loop is sufficient to account for the compensations described in the Rabbitt effect. All your PFC needs to know from its ACC foreman is that it has gotten highly competitive out there in the testing environment, and that last task was a little tougher than it expected, so you should pay more attention next time.

Cohen's model began correcting itself, and veered away from errors. It performed the Stroop as well as a practiced human—except when the investigators intentionally tinkered with its parameters to simulate diseaselike deficits. The neural net was working as a flexible feedback system without human programmers' deus ex machina–like hectoring. Occasionally in its processing of information the sim tripped its digital switches too quickly, yielded the incorrect answer, and the machine registered the conflict signal right after the error. And occasionally, after it made a correct answer, the scientists noted the presence of the conflict signal. Through this conflict-monitoring feedback setup, Cohen saw himself effectively "chipping away at the homunculus."

Cohen and postdoc Matt Botvinick speculated that an expert might predict a person's future behavior largely on the basis of his ACC firing patterns. A period of high ACC activity should be followed by quicker and more accurate responses; low ACC activation, the opposite. Jamming of one's ACC signals, they mused, should disrupt these strategic behaviors; a person would make many errors, behave recklessly. Problems in summoning the PFC to intervene would abound. Are people with ACC defects neurobiologically incapable of detecting they had a conflict or made a mistake, much less of doing anything to correct it? If so, might "normal" people with sensitive conflict monitors find inexplicable the "lack of control" exhibited by those who commit "errors" such as antisocial behavior and crime? So the uncomprehending person asks, "How could you keep on doing that stupid thing?" Such recrimination seems destined to be pointless if the subject has a disordered ACC/PFC system.

The Stroop-playing computer has no conscious awareness of its triumphs or failures; so do human ACC operations require consciousness? That is, must the ACC's call to "focus, focus, focus" be accompanied by an awareness that one is entering a thicket of conflict, or that one has made an error? "The quick answer," Cohen replies, "would be: not necessarily." The question may be unanswerable, in part because "conscious awareness" of error-making and corrections may be a secondary effect, "a story you're making up afterward 'to explain' how you were smart enough to correct yourself." You might say, "I did badly; I've got to do better."

In testing situations, however, the experimenter can often see these effects outside of people's consciousness. Consciousness can be "epiphenomenology," an illusion, a distortion in thinking the brain creates in order to portray its operations to us. Awareness may have an impact, but it doesn't mean the part of the system that's aware is actually driving the brain. "One nice feature of the model," Cohen states with satisfaction, "is that the ACC gives you a bit of an advantage without a hint of consciousness. Even if you haven't yet made an error, the mere presence of the conflict itself would be a good signal to your higher executive functions that you ought to adjust performance because you're likely to make an error if you leave the status quo.[9]

"That the ACC lies at the interface between the limbic and cognitive systems now makes sense," he says. "The limbic system is all about emotion, about placing motivational weight and significance on external and internal events." If the genius of the ACC is to gather information about the performance part of the system, it would also have to convey information about emotional states to the PFC—the system responsible for integrating feeling and knowledge, and driving motivational states. "The ACC," Cohen concludes, "turns out to be responding in a way we knew some part of the system had to."

2

REASON

Logic, Laughter, and Looking Within

In *American Ground: Unbuilding the World Trade Center*, writer William Langewiesche describes how two de facto leaders emerged immediately after 9/11: Ken Holden and Mike Burton. In the chaos of those first days after the attacks, everyone was feverishly improvising in response to a situation without precedent or rules. How these two men from the New York City Department of Design and Construction made strikingly sound ad hoc plans and seat-of-the-pants judgments at the epicenter of this massively pressurized disaster scene provide a stark display of the prefrontal executive in action. Note Langewiesche's descriptions:

> Holden and Burton . . . stood back-to-back inventing solutions to problems as they arose. . . .
>
> Mike Burton was efficient and to the point, and became known for making decisions fast and keeping the discussion on track; in one hour he could cover a lot of ground. . . .
>
> The earth shuddered underfoot, as structures collapsed far below. Burton did not allow it to distract him. . . . He did not waffle as others did.[1]

That Burton and Holden could invent, plan, revise plans, anticipate long-term consequences while avoiding tidal waves of distractions was key to their success in clearing away the ruins of the Twin Towers. These are mental operations involving reason, logic, inference, and focus; they take place in what some experts call the "mind's global workspace" that has its principal residence in the prefrontal cortex, that defining cerebrotype of our species. As opposed to these men, people with prefrontal deficits often contrive many plans for future operation, which they no sooner arrange than they abandon for new ones. They fail to commit to any plan, much less see it through to a course of action. With emerging brain-science technologies, researchers quite naturally began to explore the brain bases of these quintessentially human functions—reasoning, planning, focus, and inference.

The Architect's Story

Always, upon approaching the National Institutes of Health in Bethesda, Maryland, I am impressed anew at what a massive assemblage of buildings is this city of biomedical research, like something accreted by natural processes. Winding through its canyons to the heart of the complex, one arrives at the Clinical Building, a.k.a. Building 10, said to be the largest brick building in the world, and the world's largest hospital. At least it was ten years ago; since then Building 10 has grown another third its size again. In reality, Building 10 houses two parallel worlds. One is a clean, illuminated, and brightly painted series of spaces, where the taxpaying citizen comes for treatment or consultation. In the other world, the fifty-year-old Clinical Building shows its age with creaking doors that sometimes don't open at all, watermarked walls, and cramped spaces. There have been no makeovers in these warrens of tiny rooms that constitute the sweatshops where many scientists work.

Room 5D51 is in Building 10's second world. In the 1990s, it was not a place you frequented if you didn't work there. Certainly, neurological patients under observation did not hang out in what one denizen of the lab called "a very dismal environment." Room 5D51, at best twenty feet by twenty feet, roughly the size of a modest suburban living room, housed six postdocs in neuroscience, four research assistants, and a floating number of grad students, a crew that expanded up to more than sixteen people in the summer, plus lab equipment, filing cabinets, desks, and chairs. Most of the scientists spent ten hours a day there.

Yet it was into the dingy 5D51 that cognitive neuroscientists ushered two architects and a lawyer. One of the architects, who went by the initials PF on all the research documents, was a patient participating in a strange experiment. The other two men were healthy volunteers for the test. In 1985, PF had been diagnosed with a meningioma, a brain tumor. After undergoing the necessary neurosurgery and radiation treatments, he was said to have "recovered." But imaging scans showed a dark void in an area encompassing part of his frontal lobes. The cancer had taken out much of his right prefrontal cortex, the area directly behind his forehead on the right side.

As a student, PF had scored in the ninety-eighth percentile on his GREs in math and science. He had graduated from Yale School of Architecture, one of the elite in the country, and in the following years built up a substantial design practice. When he fell sick, he was working on his dream project, a luxury resort community in Spain. Even after the surgery to remove his tumor, he scored in the superior range on memory tests and was thus considered to have an exceptional neurophysiological profile. PF's ability to draw was intact; he could access a rich and sophisticated base of expert knowledge about architecture. Nonetheless, his life was shattered. His marriage had collapsed. He had retired from the project in Spain, was jobless most of the time, and lived at home with his mother. And he had lost his gift for design. His prefrontal functions were obviously, but subtly, out of whack.

When he came to the NIH, PF was a fifty-seven-year-old tall, handsome man of slim, athletic build, with gray hair and classic taste in clothes. Charming to talk to, with a rich vocabulary, he exuded intellect and culture. Still, there were odd things about him. He'd slip irrelevant and inappropriate remarks into a conversation, and was sometimes hypergraphically detailed about his sex life and about religion. This was part of the disorder, seen in certain frontal lobe patients, it was said, not part of his "personality."

PF came to Room 5D51 to participate in a study designed to probe the role of the prefrontal cortex in high-order mental processes. Indeed, what could be more paradigmatic of complex thinking skills than an architect's process of planning and design? And PF, more than any other frontal lobe patient the scientists had seen, was engaged in his own case, to the point where he had become quasi-expert on his pathology. It is not unusual for energetic and intelligent patients to seek out information about their disorder. But PF's involvement bordered on obsession. He was so challenging that one researcher diligently reviewed the neuroscientific literature before every meeting with him, because the architect would surely grill him about new findings the moment a relevant paper was published.

The test called for PF and two other men to redesign the dysfunctional space of Room 5D51, render the drawings, and do it within two hours. Senior neuroscientist Jordan Grafman and his young colleague Vinod Goel had custom-crafted this study around PF's special skills and disabilities as a way to surmount the "mass production" results that were typically the outcome of conventional lab tests given to people with prefrontal damage. The usual examinations rarely told investigators anything specific about what the patient's thinking deficits were in complex cases, and thus revealed little about how the PFC operates in real life.

For students of the prefrontal cortex, one of the many tantalizing mysteries is why some prefrontal patients perform normally, even scoring in the gifted range, on IQ tests and psychometric exams developed specifically to pinpoint cognitive deficits. But to Goel, these "well-structured" tests, as he calls them not a little disparagingly, measure only what someone can do in a lab situation. Any quiz with simple manipulations and a specific answer qualifies as a well-structured problem. A crossword puzzle, with only one series of right answers, for instance, is a well-structured test. And many clinical analyses of cognitive ability are constructed with even more rigid directions or sets of infrastructures the testee must follow than crossword puzzles. Some PFC-damaged people ace these tests.

Outside the lab, in what Goel called the "ill-structured" problem spaces of real life, however, the same prefrontally injured individuals—bright, charming, literate as they might be—become abject failures. Why are these people so wretched now at what they had done so well before? What capabilities have they lost from the prefrontal cortex that makes the tangled, contingent, open-ended, intrinsically ill-structured quotidian human condition so nonnegotiable for them?

The Goel-Grafman test was simple. PF noted that it was not unlike the quick-shot undergrad architectural design skits at Yale. The second architect, fifty-four, agreed it was easy enough. Serving as a healthy control subject, this professor of architecture at the University of Maryland would also redesign Room 5D51, as would a fifty-five-year-old lawyer whom the investigators had set up to act as a "novice knowledge base," a kind of benchmark of ignorance; a healthy dolt, really. (To be fair, though, imagine a test where an architect is given two hours, cold, to compose the closing arguments in, say, a corporate antitrust suit.) It is not often that a neuroscientist gets a practicing lawyer to be a guinea pig either, but the attorney had volunteered because a family member of his suffered from a neurological impairment and he wanted to help any way he could.[2]

Beyond space and time specifications, beyond the rules of architectural technique, insight, and strategy, the problem had no boundaries, no a priori right or wrong answers. Open-ended, it was an "ill-structured" problem. In deliberately vague instructions, Goel asked only that the redesign of the lab "increase our comfort and productivity." (Cleverly, the scientists hoped the test's product would give them a professionally redesigned lab.) The men were to talk out loud, to "vocalize the fragments of thoughts and ideas" while they worked, and they would be videotaped. Other rules were: you may spent up to fifteen minutes of the first hour in the lab space; you may measure, make notes and sketches, and ask anyone any questions you think relevant; you may revisit the lab for ten minutes anytime during the second hour. Please begin.

This was not an experiment most brain scientists would devise. But Vinod Goel's background was a little eccentric. When he was contemplating spending his life studying the biology of rational thought he was warned by an adviser: "The frontal lobes are the black holes of neuropsychology. You go in there, you may never return." Yet go he did, and in the early 1990s he was ahead of the curve in studying the PFC. In designing the architect's experiment Goel drew upon an eclectic perspective. Born in India and having grown up in a small Canadian town, Guelph, near Toronto, he, like PF, had a degree in architecture. "But I just cannot draw," he confessed in his office at York University in Toronto, where he is a professor of psychology. Goel had also studied philosophy, artificial intelligence, and design of intelligent machines at Carnegie Mellon and UC Berkeley, and was fluent in the language of information theory pioneered by people like the Carnegie Mellon AI founder and polymath Herb Simon, whose classes he attended. Only after all this did Goel radically change direction and study neuropsychology.

One of the initial appeals of the computational revolution in psychology, the brain as machine, was its promise of liberating us from neurophysiology, says Goel, a husky man who looks younger than his forty-six years. "But that liberation—in terms of the lack of physical constraints on computational theories—has become problematic. The constraints of physiology are powerful and must be reflected in our theories." It was the interplay of constraints of living in the real world, the ill-structured experience, against the living brain's sublime ability to plan, organize, and act, that ultimately challenged him. So by way of machine intelligence he found himself circling back to confront the messy conundrum of human intelligence.

So, untrained in neuroscience and psychology when he came to the NIH as a postdoc at age thirty-three, Goel was amazed when he surveyed the literature on mental function testing, and what results were being claimed. "If I'd been trained in that literature, grew up with it, then maybe it would have made sense. But if you looked at it as a novice would, as I was, you'd say, 'This is crazy! This is not problem-solving; this has nothing to do with planning.' It was very obvious to me then, and I guess I've been hammering away at it for over ten years now."

One "well-structured" test often used to assess prefrontal function, and one that Goel likes to inveigh against, is the Tower of Hanoi. Actually it's a moderately engrossing game, reproduced countless times in virtual form on the Internet, and a favorite of AI student programmers. In it there are three towers with a stack of disks on the first tower arranged with the smallest on top and the biggest on the bottom. The object is to move the disks from Tower 1 to Tower 3, one at a time with a minimum of moves. The hitch is that you can never put a bigger disk on top of a smaller one. Yes, the Tower of Hanoi can evaluate planning, in that one solves the problem by mentally trying out sequences of moves before executing them. And some prefrontal patients have trouble with it. But the game confounds planning skills with other irrelevant issues.[3]

And some prefrontal patients like PF succeed at it. The shortcoming of the Tower of Hanoi and other well-structured tests is that its rules and goal are completely specified. The mental transformations required to achieve this goal, while they might be difficult, are also strictly laid out. Goel likes to counterpose the Tower of Hanoi to the planning of a dinner party—an ill-structured problem. The beginning is ambiguous: How many? How hungry? How much time do I want expend? The goal, too, is open-ended: How much do I want to impress the guests? Should it be salmon or barbecue? Three or four courses? Nor are the mental transformations set in stone: Should it be catered or should I prepare it myself? Use fresh or frozen salmon? What is the priority for putting together all this?

Although most cognitive neuroscientists had never heard of "open-constraint" problems and were skeptical of Goel's plan, ill-structured problems were old news to high-level computer programmers and a special preoccupation of AI jockeys and information theorists since the 1960s, exemplified in the AI guru, game theorist, and psychologist Walter Reitman's classifications of problems. In posing any open-ended problem for their computational models, early designers of machine intelligence invariably came face-to-face with conceptual abysses like the dinner party question. They realized they had to design for the inevitably contingent, the what-if—or be forced to do the unacceptable, as Goel says, "to assume A, assume B, assume C . . ."

So to design an "ill-structured" planning problem for PF, Goel fused elements from information theory, AI, and cognitive neuropsychology. Central to the experiment was a data scaffolding, or coding scheme, he devised to analyze objectively what the three men were doing as they went about measuring, sketching, and designing. "With ill-structured tests, there is no structure in the task per se," explains Goel. The structure is imposed after the fact. You create it with your coding scheme." Goel charted each level of the scheme and digitally codified the architects' elaborations after he filmed the men on video. What they reported verbally about their design progress was factored into the coding scheme. Goel organized the designers' elaborations into units of varying size from small single snapshots of their thinking of a few seconds' duration to larger-scale operations that spanned many minutes, such as the mental interplay between the inherent problem posed by the cramped space and the skills and strategies the architects brought to it.

At the heart of the design operations are what Goel calls "transformations," the protean flow of creative purposeful thought. Most important was the "lateral transformation," the metamorphic process where a person modifies one idea into another related but distinctly different invention—as opposed to making a more refined version of the first idea. This refinement, or within-category iteration, Goel calls a "vertical" transformation.

Lateral transformations, on the other hand, involve making mental leaps or fusing disparate things into a new category. Goel often uses jokes as examples of simple, ground-level lateral transformations and has intensely examined the neural underpinning of jokes and humor. In another test, he asks subjects to imagine a piece of furniture that is a chair. Then, he says, imagine a chair that is a pineapple, or an airplane that is a fish. Subjects were to picture these conjoined images while in an MRI scanner. And when they could fuse the images, Goel saw distinct activity in their right prefrontal cortices. Vertical transforms of chair image, on the other hand, might be something like a rocker or a Queen Anne's chair and so on. And picturing a series of chair types did not require the services of the PFC.

Lateral transformations, then, may be aspects of what one recognizes as transcendental in art. And the satisfying experience of making lateral transformations may lead someone to immerse himself or herself in art. The preliminary sketches of Leonardo da Vinci, Monet's garden paintings, Bach's *Goldberg Variations*, or Beethoven's *33 Variations on a Waltz by Diabelli* are perhaps extraordinary celebrations of overt lateral transformations.

When the two hours were up and the men turned in their work, it was in the phases of lateral transformation that the two architects' efforts were most radically different. The healthy architect delivered a clearly reworked lab plan with an airy, improved circulation pattern, more open space, and better lighting. He established a coherent territorial spatial hierarchy among the senior scientists' workbenches and part-time technicians' areas. He offered finished and blueprint-ready detailing in furniture, workstations, doors, corridors, and more. He provided an elevation sketch and a cross section. Goel could clearly see how the healthy architect efficiently defined the parameters of the problem, then how, moving on to the design itself, he spent most of his time working on his plan in its several incarnations. In his mind, the architect juggled information about the lab space, its personnel, their work needs and behavior, and then held all this information in mind through the preliminary sketches, the emerging design, all the way to the refinement details at the finale. He had "solved" the problem, transforming an inefficient, ugly workspace into one where workers' physical and emotional comfort and productivity were enhanced.

PF's design told a story of disarray. It was unfinished, the plans were mere scrawled fragments, one on top of the other. PF allocated his time differently, too. He spent a greater percentage of the two hours trying to clarify the problem per se. Then he made several attempts to generate ideas for a plan but was unable to develop and explore any vision through lateral transformations, the "Try A. Reject that. Try plan B. Return to A and fuse it with C" iterations. Each of his idea fragments appeared unrelated and independent.

Yet even in the midst of this mental chaos, PF was able to articulate his disorder: "I know what I want to draw, but I just can't do it. It's crazy . . . Even as a student, there would be sketches on top of sketches . . . It would be progressive," he ruminated in his videotaped commentary. "Here I seem to be doing several different thoughts on the same piece of paper in the same place . . . It's confusing me. So instead of the one direction that I had in the beginning, I have three or four contradictory directions with not a kind of anchor to work from . . . It's as if I'm getting a train of thought, and then I start to draw it and then I lose it. Then I have another train of thought that's in a different direction, and the two don't—" His thoughts stopped cold; the actor was portraying his own prefrontal train wreck.

"He knew what these things were and what he should do, but couldn't execute them," Goel says somberly. "PF's drawing skills were there, his memory was intact. But in terms of designing this simple thing he just didn't know which direction to go. He couldn't say, 'I'm going to do this,' then go away, and come back with it done, hang on to it, then move on to the next step. He could not hold in mind the big picture of the problem while manipulating the parts; could not develop a strategy or structure for moving forward." PF never resumed design work, but he sometimes got jobs doing "as-built" drawings, that is, rendering sets of blueprints and documents with up-to-date changes and additions to the original blueprints of a building. It was well-structured work. He died five years after the redesign of Room 5D51.

And what about the lawyer? His creative session lasted only twenty-five minutes, which Goel brought to a merciful end. Because his drawing skills were limited, to put it charitably, he made paper cutouts of the desks and equipment. And, typical of a novice, he worked on a concrete superficial level: arranging furniture. He never considered the jobs of the lab workers or their social concerns, nor pondered the infrastructural problems that preoccupied both trained architects. "He wasn't aware of the fact he couldn't do it," says Goel. "He'd think, 'Put a desk here, put a chair there, and that's it.' There was no sense that this was a hard problem. He didn't say, 'I can't do it; it's too frustrating.' He just knocked the problem off. 'Here's the solution; that's it.' When you compare the process of a professional and a novice it's clear a novice is working on a very superficial level, but he may not be aware of that."

The lawyer did, however, engage in several lateral transformations and develop his "plan," such as it was. And he did hold a representation of his concept in mind while he explored several configurations of it, moving around the cutouts within an outline of the lab space. In other words, the lawyer used his noggin, his basic prefrontal operations. His almost laughable failure, the researchers concluded, was simply the result of his impoverished knowledge base. And thus his effort was consigned to a footnote in the final paper, a kind of a comic coda buried beneath the misfortune of lost talent and the scientific hunger to explore the biological bases of thinking in the PFC.

Goel has conducted a number of a large-scale ill-structured-problem experiments, including one with nearly eighty prefrontal patients. This scenario is a fairly diabolical long-range planning scheme—family travel plans. Here his patients must work out for a hypothetical Toronto family of four a first-time, weeklong vacation to Italy. "They have X amount of money," Goel outlines, "they have a specific set of interests, they don't speak Italian—and they have to plan a trip." Goel has several variations on this scenario, using a family who speaks Italian, for example, to test for the impact of this knowledge. So far, he says, his coding scheme is working well.

In Goel's ill-structured experiments, nothing is more central than the creation of the mental "problem space," a term adopted from the silicon vocabulary, and from artificial intelligence expert Allen Newell's great artificial intelligence SOAR project. In these AI programs, inflowing perceptions, working-memory units, and long-term knowledge combine to generate a model of the world circumscribed by the problem—its sensations, facts, relationships, and the actions intended upon it. To simplify matters, one might envision a problem space as an ad hoc online chat room, or a multidimensional Shakespearean play-within-a-play, where, within the theater of the mind, a more circumscribed stage show is performed.

A vast amount of information processing takes place in our mental workshops—much of it unconsciously. Here we are free to experiment with the consequences of various actions, inferences, and reiterated representations, which we revise into new representations of the plan, some of which will appear good enough to try out in the "outside" world. The genius of the prefrontal cortex is that can create SimWorld applications from any script that comes its way. How a problem is solved—what seems easy or difficult—no doubt depends on how your PFC organizes and executes all the disparate neural computations involved.

Traditionally, frontal lobe patients are said to have planning deficits. "What I didn't fully appreciate at the time," Goel reflects, "is that the patients' difficulties are not so much in planning, but in creating the structure that allows planning to happen." Before a messy, real-life problem can be solved, some kind of structure must be imposed on it. A person must be able to gather the information he needs to create his problem space out of background knowledge, details of the scenario, extraneous sources of facts and advice in the world, and so on. And then he must organize, transform, and "play with it," going back for more details when required, discarding items that lead to dead ends. Only then can a coherent plan emerge. Using the play-within-a-play conceit, a person sets up various scripts and sees how the dramas play out.

To be extraordinarily talented at performing prefrontal problem-structuring operations may constitute elements of what we call "genius" in today's overused sense of the word. Take the oft-cited football "genius" of New England Patriots coach Bill Belichick, extolled for his in-game plan execution. During an interview on *The NFL Today*, Patriots quarterback Tom Brady said this about Belichick: "He has a tremendous eye for detail, and he's constantly anticipating things . . . He's also an incredible tactician. He keeps modifying the plan as the details change." Interviewer Bob Costas asked, "So he's got a plan B, if A doesn't work out?" Brady: "Plan B? No, he's got plan C, D, E, F, and so on ready before A doesn't work out. But all those are subject to change as the game unfolds."

A plan represents a blueprint for achieving some future condition, and although its rightness cannot be fully known until it is actually executed, a planner needs some self-measure of its solidness as he develops this solution. This feedback originates in one's mental workshop, from which viewpoint one evaluates and tries out schemes and variations on schemes. The lack of internal feedback in patients with prefrontal damage tends to keep them stuck in the present. Without the ability to be an effective director of their dramas of tomorrow and next year, the future for them is increasingly unreal.

Besides problem-structuring, frontal lobe patients also have trouble with retailoring solutions to fit changes in reality. They are sometimes unable to take advantage of the fact that constraints on real-world problems are negotiable—that a housewife, for instance, could go out and get a job, or that someone has the option to quit his job and set up a computer consultation company at home. Healthy people are much more likely to consider these open-ended possibilities. This stuck-in-a-rut perspective is consistent with frontal lobe patients' impaired ability to shift among multiple mental sets. Frontal lobe patients are also swifter to determine that they have satisfied a problem's requirements. Again this determination requires a self-referential evaluation, and they tend to believe they've created a complete series of plans, when in fact they have not.

So Goel sees that there is "no single unifying difficulty patients with frontal lobe lesions encounter that can be termed a 'planning' deficit." These are multiple deficits, and they will affect one's ability to negotiate many real-world situations where plans and goals may or may not be embedded in the structure necessary to successfully carry out complex mental calculations. Goel's earlier patient groups were not limited to people with lesions in the same areas of the PFC, so little information could be gleaned about structure/function relationships within prefrontal subcomponents. But within the terra incognita of the neural substrates of rational thinking and problem-solving, Goel searches to understand prefrontal function in reasoning—the bedrock of planning and foresight.

At the core of reasoning is logical thinking. And logic is very seldom a pure process but is confounded by emotions and "rituals of
mind," for lack of a better word. The media frenzy surrounding the
unfortunate Terri Schiavo in 2005, for example, dramatically highlights the ways belief systems and logical thinking can come into
conflict. For a minority of onlookers, the language of Schiavo's
mother begging the Bush administration to "save my little girl"
seemed a more compelling and powerful message than the logical
deduction that Terri was, in fact, beyond saving.[4]

Investigating the neural substrates of reasoning and the interplay of logic and belief is new in brain science. Not surprisingly, the
prefrontal cortex is at the center of a nexus where logical thought—
and more often, logic freighted with belief and bias—commingles
with emotion. Vinod Goel is one of the first to investigate how this
all is interrelated. How, he wondered, could logical thought be isolated and tested in the lab? Of logic's various forms, he thought,
deductive reasoning—being the most specialized, narrow-focused—
might lend itself best to precise research purposes.

All Men Are Mortal; Socrates Is a Man; Socrates Is Mortal.

In everyday life, as in philosophy, we dissect the premises of logical
arguments to see if they are durable enough to support their conclusions. Deduction is a closed loop of logic, requiring no external
input for its operations. The "truth" of a deductive argument, such
as the Socrates syllogism above, relies on the claim that, subsumed
within its premises are absolute grounds for the conclusion. The
truth of the premise (all men are mortal) guarantees the truth of
the conclusion (Socrates is mortal). If you buy the premise, you buy
the conclusion. The conclusion of a deductive argument, moreover,
can be independent of the content of its premises. That is, the conclusion of the syllogism is valid whether the content of the premise
is Socrates or Joe Blow. Indeed, one can substitute "hedgehog(s)"
for the terms "man" and "men" and the conclusion will hold.

For the past twenty years, two theories vied to explain how the
brain computes deductive reasoning. One posited that deduction is
underwritten by a language-based system. That is, our mental model
of a deductive argument preserves elements of the linguistic struc-

ture in which the premises are stated. The second hypothesis insists that deductive reasoning marshals a visual-spatial neural network, to build in one's mind a pictorial model of the problem. Here the mental representation preserves the structural properties of the world and its spatial relationships that the deductive problem is about.

During decades of heated debate, there was no way to test the predictions implicit in either theory. So given the chance in the 1990s, when imaging technology began to ask cognitive theory to put up or shut up, Goel put the two postulates to a test. In his initial study, he PET-scanned the brains of volunteers while they assessed the truth of a simple Aristotelian syllogism: *All apples are red; All red things are sweet; All apples are sweet.* As the participants pondered, Goel saw clear evidence of activity in a left-hemisphere network that included the language-processing areas of the lower left frontal lobe, the lateral PFC, and the mental libraries of the temporal lobe. He detected no activity in the right hemisphere, the putative dominant hemisphere for spatial processing.

This seemed to validate the linguistic model. But Goel was not satisfied, especially since many people avow they think through logical problems by constructing picturelike images or even 3-D models in their minds. The "All apples are red" syllogism was concrete, word-based. What would happen if the syllogism was explicitly spatial— about the relationships of objects in different spaces? So he repeated the experiment with the argument *The apples are in the barrel; The barrel is in the barn; The apples are in the barn.* Again he found only left-hemisphere activity—including areas of the dorsolateral PFC, the anterior cingulated cortex (ACC), and the superior temporal lobe. He found nothing going on in the right hemisphere. Nonetheless, the syllogism was still word-based, thus "verbal," so Goel went back to the drawing board.

By 2000, and now using the more dynamic fMRI technology, Goel deployed a syllogism formally the same as the "apple" sets but stripped of verbal, semantic content: *All A are B; All B are C; All A are C.* The subjects could not tap their language-processing brain areas for this one. Although again Goel found the left-hemisphere linguistic network fired up, there was something new. Spatial-processing areas in the parietal lobe in the left hemisphere also lit up, along with the bilateral PFC (BA 44, 8, 9). But the temporal lobes were quiet. The abstract syllogism, then, engaged a different circuitry—

PFC-parietal, rather than PFC-temporal—than did the concrete, word-based syllogism.[5]

Because this network used parietal areas recruited for abstract reasoning involving math and for processing spatial information, the finding suggested that when contemplating abstract and arbitrary—rather than contentful—deductive problems, people do indeed build spatial models ("picturing," say, the relationships between A, B, and C). In these nonverbal syllogisms, then, the evidence of brain activity affirmed the involvement of visuospatial systems. The more abstract the problem, perhaps, the more you need to create an internal workshop model. Language logic, on the other hand, ties to individual memory of experience: that is my knowledge of apples, barns, barrels, and so on.

Wanting to observe the brain grappling with pretzel-like kinks of deduction, Goel used trick syllogisms such as *All pets are poodles; All poodles are vicious; All pets are vicious.* Here, the "truth"—or lack thereof—in the content contradicted and vied with the syllogism's framework of logic. And he saw regions of the right PFC light up. As did the ACC—not surprisingly, given the conflict arising from wrestling with the dissonance between the syllogism's logical rigor and its real-life wrongness. (Indeed, in serving up conflict, these syllogisms constituted a kind of logical Stroop test.) The right-hemisphere network, Goel suspected, was turned on by the presence of something extraordinary, weird. "The right frontal hemisphere is engaged in unusual as well as contradictory reasoning situations," he surmises. "Take: *All apples are red; All red things are poisonous; All apples are poisonous.* You say, 'Yes, but I know all apples are not poisonous.' There's a contradiction between logic and belief." To resolve that conflict, you must make the appropriate validity judgment about the "nonbelievable" conclusion. To do that, you bring in the right PFC.

Same with meaningless content. If one reads: *All blims are blue; All blue functions are gleets; All blims are gleets*, the formal logic is identical to *apples are red*, and yet we find right-PFC activation for *blims are gleets* that is not there for *apples are red*. As soon as we introduce a syllogism with no real-life content, or content we are not familiar with, or content that displays a contradiction between logic and belief, then the right PFC system is deployed.

The right PFC network, Goel thinks, is more attuned to reasoning that is nonconceptual, contradictory, or incoherent. The left-hemisphere system, meanwhile, appears more specialized for handling linguistic, "reality-based," factual material, and thus dominates when we reason about familiar problems and scenarios. Most people are right-hand/left-hemisphere dominant. Humans are creatures of language; we are conceptualizers. Conceptualization, then, may be our default mode, so the left hemisphere becomes the necessary and often sufficient processor of everyday reasoning. The right hemisphere's resources may jump into the fray to confront a logic beyond the quotidian, to go one-on-one with something that challenges our sense of reality. The processing equipment of the right PFC, coupled with that of its parietal lobe partner, facilitates the building of mental models in puzzling, nonconceptual, or incoherent situations.

Goel's findings mirror the thinking glitches of some frontal patients. In deductive logic tasks, people with left frontal injuries are more basically impaired than those with right frontal lesions. Right-hemisphere patients, however, have more difficulty in reasoning with abstract, contradictory, or incoherent material. In an undemanding cognitive test, patients were asked: Mike is taller than George: who is taller? Left-hemisphere patients had difficulty with these bedrock-simple relational problems. But right-hemisphere patients only stumbled when the form of the question was incongruent with the premise, for example: Who is shorter?

Although there is scant research on the neural substrates of rational thought, another group sought to replicate Goel's findings using more finely honed arguments. Would the most stripped-down reasoning problems, such as *If/then* statements, excite the left PFC-parietal system? Ira Noveck at the Institut des Sciences Cognitives in Lyon, France, with Goel and others used two kinds of reasoning statements while they scanned volunteers. The easy sets were straightforward *If A/then C* statements: *If it is raining, the sidewalk is wet/the sidewalk is wet/then it is raining*. The harder versions have an *If A/then C* major premise but with a minor premise: *not C*, which justifies the conclusion *not A*. For example, when we are informed that *If it is raining then the sidewalk is wet*, and consequently told: *the sidewalk is not wet*, we logically conclude: *it is not raining*. This strategy Noveck called *reductio ad absurdum*.[6]

Healthy people usually get easy conditional syllogisms quickly and about 90 percent correctly under testing conditions. The reductio syllogisms take longer to process and are correctly answered about 60 percent of the time. Noveck used abstract versions of the same kinds of syllogisms as Goel's "contentless" letter syllogisms. When the subjects processed easy syllogisms based solely on letters, they engaged the same left parietal-PFC circuitry activated on Goel's more conflicted syllogisms. The harder versions fired up the same system, plus the anterior cingulate. This supports the notion that basic reasoning with abstract materials requires the left PFC-parietal network. Harder tasks stimulate the parietal regions to fire more intensely, Noveck surmises, because the reductio ad absurdum syllogisms require greater computing power, first to make a supposition, then to see that it leads to a contradiction, and finally to reject that supposition. Plus, all of this information processing must be temporarily stored in the PFC's working-memory units.

Why the does the parietal lobe light up for abstract stuff, and the temporal lobe for meaningful content? The left PFC/temporal lobe network is probably the default logic system, because problems that are meaningful in real life more closely resemble conversational exchanges, and thus prompt the thinker to engage in a wider, everyday range of inferences—also prompting a trip to the temporal lobe storehouses. Deductive thinking about actual things, as well as beliefs, in an experienced world resembles verbal communications we've had with other humans about those things. Left PFC/parietal activity, on the other hand, may work as part of a more general-purpose system engaged when you have fewer "facts"; the abstract is in a sense isolated from immediate reality (for example, $E = mc^2$).

Mental models, then, enable the PFC to grapple with and manipulate abstract or confusing elements to confer validity (or not) upon given logic. In 2005, one might say, some Americans were running these conflicting computations—confronted as they were in the Terry Schiavo case with an instance of cognitive dissonance. Here for some people, logical biomedical evidence powerfully clashed with faith and emotional attachment to the belief in life forces that transcend the vitality of the cerebral cortex itself. Some people may have let belief systems overwhelm logical processing. To paraphrase William James, a great many people think they are thinking when they are merely rearranging their beliefs.

Goel began probing what goes on in neural pathways when "belief-bias" confronts logical thinking. Within deduction's closed system, how do people harness—or suppress—their beliefs about the world to influence how they conduct logical thought? Within the logical scaffolding of pure deductive reasoning, beliefs should be irrelevant. So the recurring phenomenon of belief-bias was puzzling to him. In the *All apples are poisonous* conclusion, for example, if we insist this is not true, our belief about apples overrides the deductive paradigm.

Not surprisingly, we reason better when the truth of a conclusion coincides not only with the logical relationship between premises and conclusion but also with our beliefs about the world. In such cases, beliefs are "facilitory" to logical reasoning processes. Note this syllogism: *No cigarettes are inexpensive; Some addictive things are inexpensive; Some addictive things are not cigarettes.* Test subjects deemed this argument valid 96 percent of the time. Yet on an argument having the same logical form but with an unbelievable conclusion—*No addictive things are inexpensive; Some cigarettes are inexpensive; Some cigarettes are not addictive*—people accepted it as logically valid only 46 percent of the time. If the deductive conclusion is inconsistent with our beliefs about the world, our beliefs can inhibit our application of the logic.

To see how belief-biases change the neurobiology of logical reasoning, Goel, with the British imaging expert Raymond Dolan, bade volunteers to confront a mind-boggling 120 syllogisms organized into categories in which the levels of "belief-truth" or "belief-falsity," "logic-validity" or "logic-invalidity," and neutral controls were mixed. The participants had to judge each syllogism on merit of its logic alone. Whether or not it was "true" or "believable" was not relevant. Here is an example of a "belief-neutral/invalid" syllogism: *Some monorchids are ground rhumbs; All ground rhumbs are rare; Some monorchids are not rare.* Most participants had zero beliefs about monorchids or ground rhumbs and could see clearly that the logical structure of the syllogism was fallacious.[7]

Among belief-laden samples, Goel included both valid and invalid, true and false syllogisms. Here's an example of a belief-true but logically invalid syllogism: *No reptiles can grow hair; Some elephants can grow hair; No elephants are reptiles.* Here's a belief-false but logically valid example: *Some green amphibians are toads; All green amphibians are frogs; Some frogs are toads.* And finally, a belief-false and invalid syllogism: *No unhealthy foods have cholesterol; Some unhealthy foods are fried; No fried foods have cholesterol.*

The right lateral PFC and parietal system, it seemed, came on strong when the participants suppressed a strong belief-driven impulse to avoid making an incorrect judgment about a syllogism's logical structure and when they forced themselves to correctly state its logical validity or invalidity; that is, when they overrode their belief-bias to assess the syllogism solely on its logical merits. This was consistent with the right PFC's role in cognitive function. By contrast, when a subject's logical reasoning was bested by his or her belief-bias, the scientists saw another frontal area, the ventromedial prefrontal cortex, light up. This was the first neurobiological evidence for a two-track reasoning system within the PFC itself. On one track, the "higher" lateral PFC network turned on to override a belief-based process; on the other, a "lower" ventromedial PFC area outweighed logic and enabled a belief-bias to prevail.

The ventromedial PFC (sometimes considered the medial OFC, or number 6 in figure 1 on page ix) is deeply tied to emotional processing. Located against the inner walls of the PFC and surrounded by the orbitofrontal cortex, the ventromedial PFC plays a big role in evaluating reward and has powerful two-way links to subcortical limbic centers. Ventromedial involvement suggested to Goel that effects of belief-bias in reasoning are influenced by emotional processes. Goel's findings are in sync with those of the Princeton psychologist and philosopher Joshua Greene, who suggests that emotionally laden "intuitions" seem to appear suddenly and effortlessly, with a tag of "good" or "bad," but without any sense of having gone through steps of weighing evidence or coming to a conclusion.

Thus the excited ventromedial PFC phenomena during belief processing pointed to the next question: how do the brain's emotional systems affect deductive reasoning? The scanning data hinted at outlines of an answer: a kind of competitive, seesaw dynamic between the lateral PFC in the right hemisphere and the ventromedial PFC in both hemispheres. When the ventromedial PFC was hopping with activity, a person was more likely to judge based on belief—even when belief produced a logically incorrect answer.

If the subject produced a logically correct answer—even when contrary to what he or she believed—the lateral PFC held sway over the ventromedial PFC. In other words, if a rational response was overridden by belief, there was greater ventromedial activity; if the rational overrode the emotional, the ventromedial PFC was suppressed by the lateral PFC. "We conjecture the right lateral PFC serves to detect and/or resolve the conflict between belief/emotion and logic. The PFC may inhibit this ventromedial prefrontal activity," Goel says. Or not. If the right lateral PFC does not preempt ventromedial activity, belief prevails. Obviously this system generally works well enough for us to distinguish the validity or invalidity of an argument in the face of some conflict with what we believe or emotionally feel about the world.

But further, there was this difference between belief and emotion to tease out. Not surprisingly, it turned out to be a complex interaction. If you say the conclusion *All apples are poisonous* in the previous example is invalid because it is untrue, your belief system has incorrectly overridden your logic system. "You know it's false," says Goel, "but nothing to jump up and down about." The lateral PFC can handle it with a little help from the anterior cingulate; it's cold reasoning. "But if I say, *All Irish are drunks* or *All Muslims are terrorists*, people take objection." That's hot. In the next experiment, Goel and Dolan collected samples of "hot" logic, calculated to upset people—including syllogisms nastily denigrating a United Nations-ful of constituencies. "We included all ethnic groups just to be fair," Goel adds. An example of a "hot" syllogism included, *Some wars are justified; All wars involve raping of women; Some raping of women is justified.* These were mixed up with emotionally cold syllogisms, such as, *Some Canadians are not children; All Canadians are people; Some people are not children.*

Despite the logical formalism of both "hot" and "cold" syllogisms, lateral and ventromedial PFC again showed yin/yang firing patterns, depending on the emotional charge or "saliency" of the content. "Cold" reasoning trials resulted in enhanced activity in the lateral PFC area, and suppression of activity in the ventromedial PFC. "Hot" reasoning trials resulted in the opposite: enhanced activation in the ventromedial PFC, and dampened lateral PFC firing. This reciprocal engagement of the lateral and the ventromedial PFC provides evidence for a dynamic reasoning system, the configuration of which is strongly influenced by emotional strength.

Where the subjects accepted "hot" logical conclusions even if invalid, the ventromedial PFC prevailed over its more "rational," "cold" lateral prefrontal partner. If we have a conflict between logic and belief, and we go with the belief—either we can't detect the conflict, or can't resolve it, and go with the belief—then the ventromedial PFC is dominant. It appears that belief and emotion make use of the same or similar neural pathways and systems; but the scientists have not yet, in fact, succeeded in teasing apart their functional differences—if they are there.

This evidence corresponds to studies of patients with ventromedial damage, who are often described as being "too rational" and unable to integrate logical with emotional responses. Thus one might conclude that people with ventromedial lesions have an advantage over normal individuals in the Platonic clarity of their logic. But in reality, they tend to make poor real-life decisions, probably because they can't "read" their own valid emotional and bodily signals—those "gut" responses, "intuitions," and "hunches." Ventromedial deficits could underlie some behaviors subsumed by the media's caricature of the "cold-blooded psychopath," as we'll discuss in chapter 3.

Ventromedial PFC activation in "hot," incorrect trials about ethnic groups implies its role in the neural substrates of ethnochauvinism and prejudice, bigotry, and racism—in the intense, erroneous, pervasive, and persistent belief-biases people hold about various groups of "others." If one's overpowering belief-bias is fairly hardwired, it would be difficult to neurally detect the fallacy in the premises of a seemingly logical but biased conclusion, or suppress the emotion that is forcing the "illogical" conclusion. The emphasis on quick, automatic emotional reactions is supported by evidence that people evaluate others and apply morally laded stereotypes automatically. People readily construct post hoc rationales to justify their judgments and actions.

How, then, are racism, prejudice, and other "isms" to be overcome in the face of these possibly hardwired neural diagrams? How, too, do the everyday currents running through the "hot," biased, ventromedial pathway affect problem-solving in general? A 2003 study suggests there is a mental price to pay for maintaining an incongruent deductive computing system. Through psychological profiling, Dartmouth's Jennifer Richeson and colleagues identified a group of thirty white males as racially biased against black people.[8] They also scanned these volunteers while they viewed pictures of black faces, and saw some of the men's dorsolateral PFCs light up. Richeson interpreted this to represent an attempt by these men to suppress racist sentiments. Next, after meeting black men face-to-face, the men took standard intelligence tests. The volunteers identified as racist fared worse on the tests than nonracist controls. It may be that that harboring racial prejudice, even unwittingly, created the need to suppress this stereotype that sprang so automatically to mind. And this effort basically siphoned energy from their dorsolateral PFCs, draining them of capacity for the intelligence test's higher-level computing needs.

One also wonders if Goel's two-track prefrontal reasoning system might help explain the powerful grip of ideologies and religious fundamentalism. It may help us understand the immutability of the "logic set" of a fanatical political believer even when he or she is confronted with overwhelming evidence to the contrary. When yoked to belief-bias, deductive logic may indeed serve as a hand-maiden to fascist thinkspeak. The ideological interior proof stands impregnable, rendering an absolutist conclusion—sustained by a "hot" emotional belief system powerful enough to withstand the scrutiny of objective inference or correspondence to reality.

In examining political behavior, John Jost, now at New York University, focuses on something called system justification theory. He seeks to understand how and why people provide "cognitive and ideological support" for the status quo, even when this support seems at loggerheads with their individual and community interests, especially in disadvantaged socioeconomic groups. This is a phenomenon resembling that described in Thomas Frank's book *What's the Matter with Kansas?*, where blue-collar and rural poor buy into a right-wing economic agenda that is in fact detrimental to them. According to Frank, they do so because these policies are yoked to hot-button moral issues such as abortion and gay rights. Jost is now investigating the psychological bases and underlying cognitive and motivational differences between liberal and conservative ideologies. In 2003, his group published a review that statistically summarized dozens of studies conducted over a half century dealing with differences associated with left- versus right-wing thinking. They found that the likelihood of adopting conservative rather than liberal political opinions was significantly correlated with a sense of threatening social instability, fear of death, intolerance of ambiguity, need for closure, and lower cognitive complexity.[9]

Discoveries about the neural substrates of belief and rationality, furthermore, raise speculations about the deeper bases of philosophy and faith. Is religion the ultimate theater for the battle between valid and invalid, true and false, deduction? Is faith processed as a closed-loop logical syllogism by the frontal-parietal network recruited for abstract deductions? Or dominated by an emotional PFC network? Or a mixture of some kind? One thinks of Graham Greene's preoccupation with the "paradox of faith," the "I believe because it is absurd," or of Kierkegaard's "teleological suspension of the ethical." The Danish existentialist brooded intensely about the conundrum of Abraham's avowed sacrifice of Isaac. Abraham had no choice but to obey the command of God, Kierkegaard thought, since there was no adequate proof, no recourse to outside logic, within faith. In the dominance of faith over logic, Abraham had to follow orders. Yet the construct of a metalogic of a faith beyond logic is itself a kind of absolute self-enclosed chain of deductive premises, followed by a conclusion validated in the rivers of passionate belief.

Ventromedial emotional activity co-opted by a higher, lateral logical process? There is no shield completely walling off belief and emotion's corresponding neural impulses from the brain's logical processors, even though the lateral PFC has great powers of inhibiting these pathways. What emotion-drawn neural operations propel some of us to construct categorical imperatives—from the single-neuron level Earl Miller observes in his macaques' dorsolateral PFCs to the system-wide activation Goel sees when one judges the truth of abstract formal logic?

The Induction Machine

An episode of *CSI: New York* titled "A Man a Mile" features the suspicious death of a construction worker in a subterranean aqueduct connecting the city to its upstate water supplies. After descending into the depths of the tunnel, Detective Mac Taylor, the Gary Sinise character, states in a signature remark, "[What we've got here] are pieces of evidence looking for the connection." In the course of the show, the CSI team puts the parts of the picture together, builds an hypothesis for the sandhog's murder, and ensnares the suspect one painstaking clue at a time. What Taylor and his forensic team practice, and what is in fact the basis of all popular mysteries since Edgar Allan Poe, is inductive reasoning.[10]

In the *First Book of Aphorisms*, in the early 1600s, Francis Bacon articulated a revolutionary method for interpreting reality: we interpret the world around us not through deductive syllogism but through induction, "rising by a gradual and unbroken ascent" from observable evidence, through intermediate laws, to the underlying principles of nature. Bacon emphasized that as interpreters of nature, humans derive knowledge and theory not from deduction but from empirical evidence and the fruit of experience. While there is something "safe" in the certainty of deduction's closed world, inductive knowledge is never absolute but always tentative, awaiting further investigation and affirmation. Yet induction's highly educated guesses allow the leaps of comprehension that drive the human race.

We use both induction and deduction, often in circular fashion, one supporting the other. For example:

Ken: I've noticed that every time I squeeze a balloon, the harder I push, the harder it seems to push back, or when I lie on an inflatable mattress, it compresses up to a point then seems to stop. So I guess this is because as I decrease the volume of a confined gas, the pressure it exerts increases.

Barbie: That's Boyle's law. The volume of a sample of gas is inversely proportional to its pressure, if temperature remains constant.

Ken is using inductive reasoning, arguing from observation, while Barbie, arguing from Boyle's law, employs deductive reasoning.[11]

Although deduction may be a "purer" starting point to explore the neurobiological bases of rational thought, induction—a more ubiquitous, expansive, open-ended, and seemingly effortless form of human reasoning—is harder to parse. "The greatest puzzle in psychology, and ultimately neuroscience, is the puzzle of induction. Induction is magic," proclaims Goel. "We do not understand the neural bases of inductive processing." Whereas in deductive arguments, no new information is added and the conclusion restates information present in the premises, the inferences and conclusions of inductive thought can catapult the thinker far beyond the information included in the premises. This is the kind of reasoning we engage in every day, and it is the essence of most problem-solving, planning, and creativity.

Psychological models generally define induction as hypothesis generation and selection: a person searches a large database to determine which pieces of information are relevant and how this information will be mapped onto the problem. For instance, Goel presents a pair of his patented syllogisms: *George is a mammoth; George eats pine cones; All mammoths eat pine cones.* This is an invalid deductive argument. Most people can accept it as invalid, though they also can accept it as a plausible or reasonable argument—despite the fact that they have no personal experience of mammoths. The next syllogism also contains an invalid argument: *George is a mammoth; George has a broken leg; All mammoths have broken legs.* Even though it has the same logical structure as the previous argument, we do not recognize this one as plausible; we say it's ridiculous. Although the property of eating pine cones probably applies for the species, the property of having broken legs does not. So what properties generalize and what do not? How do we make these judgments?[12]

In another extinct-species exemplar, Goel cites the unearthing of dinosaur fossils buried in an Alberta tar pit. Since the specimen, a tyrannosaurus, had eight-inch-long, razor-sharp teeth, one can infer that all tyrannosaurs had long, razor-sharp teeth. But we do not infer that all tyrannosaurs drowned in tar pits. "Somehow" we recognize that tooth size is a relevant property for generalization across species, while the mode of death is an individual accident. In both the mammoth and tyrannosaurus cases, we make inferences. The puzzle of induction is to a large extent, Goel thinks, the question of how we make these judgments of relevance. He says, "The central problem of inductive reasoning is to determine which properties are relevant and which are not."

To explore the neuroanatomical bases of induction, Goel and Dolan scanned volunteers as they categorized sets of imaginary animals. Named "Caminacules," after their creator J. H. Camin, the hypothetical creatures resembled tiny pond-dwelling nematodes and diatoms, with multiple leg groupings and feelers. In a two-part task, the participants were required to determine if all critters in a set were of the same "species." In the easier version, the subjects

were given a rule for classifying Caminacules: if the animals displayed the same-shaped tail and abdomen, then they were the same
species. In the harder version, the subjects had to infer their own
rules for determining the animals' species.

Here the participants tended to try out routine visual strategies,
comparing the Caminacules to known bug and animal types; these
tactics invariably failed, since the Caminacules were new to them.
Next, they fell back on analyzing features such as abdomen shape,
arm and leg numbers, horn and feeler sizes, attempting to see which
body parts were relevant for classification purposes. "As you went
from easy to hard, there was the element of the unfamiliar. How do
you conceptualize what you haven't seen before?" Goel asks.

The easy task activated a network including bilateral parietal and
prefrontal areas, especially in the left hemisphere. The easy inference problem also activated the hippocampus in both hemispheres.
Goel interpreted this to indicate that the hippocampus was engaged
in encoding "species rule" data in memory banks. Since the participants could not incorporate the hypothetical beasts into preexisting categories, they needed to anticipate and prepare a response to
the rule stored in, then accessed from, these memory stores. Then
working-memory areas of the lateral PFC held the rule online while
the brain network carried out feature-by-feature searches.

In the harder task, people could not anticipate the rule because
it was unknown. With no memory systems coming into play, there
was no hippocampal activity. Subjects did have access to knowledge
databases for categorizing animals in general: features of
appendages, body markings and shapes, positioning of eyes, and so
on. When they used these general systems of hypothesis generation
on the Caminacules, lateral PFC areas fired, but more intensely in
the right hemisphere. The predominance of right PFC activation in
the hard inference task was consistent with its firing during contradictory, incoherent, or abstract arguments in deductive reasoning
tasks. But surprisingly, also active in the hard inductive task was an
area in the right orbitofrontal cortex. How might this area contribute to the special hypothesis-selection and category-creating
work needed for inferring?

The orbitofrontal cortex (OFC) (see figure 1 on page ix) is crucial in linking emotional brain zones to more rational ones. The
OFC seems to be centrally involved in mental operations involving

motivation, and in evaluations of reward choices—judgments such as, "Is this payoff big enough to justify that action?"—the whole spectrum of means versus ends. The orbitofrontal cortex is an adjustor, calibrating one's behavior to shifts in rules and contingencies. (When a traffic light at an intersection freezes, say, motorists start creating ad hoc strategies to decide when to stop or go.) This PFC sector comes online when there is insufficient information available for us to easily determine the right course of action. It's recruited in "what should I do now!?!" times where you reach the best solution by weighing the nuances and values of suboptions. The OFC may operate when certainty cannot be achieved and ambiguity cannot be eliminated; when the second-best solution is all that you can hope for.[13]

Extrapolating from these OFC functions, then, one might "infer" that inductive reasoning does indeed call for the OFC's special talents to gain the "reward" of figuring out how to classify the Caminacule. This lab exercise is, after all, a miniature version of the inductive process taxonomists use to identify and classify new organisms, not to mention Darwin's monumental works of induction. Humans find inductive thinking riveting drama and great fun as well—witness our love of sleuthing, making connections. Our brains revel in it.

In his search for the neural substrates of incentive, Adrian Owen, too, wanted to understand how objects and events in the world acquire intense motivational value and drive us to certain behaviors, even in the absence of a clear biological need. In an imaging study using restaurant menus, Owen saw the OFC light up when participants mulled over several favorite dishes as they selected from a series of possible ones. The lateral orbitofrontal cortex selectively lit up when the prospective diners had to suppress responses to many mouthwatering dishes to select the ones they wanted most. This again suggests orbitofrontal involvement when one tries to infer the best choice from an array while rejecting the others.[14]

There are anecdotal stories about a patient with OFC damage who planned to dine out, but spent the evening driving from diner to café to four-star restaurant, unable to pick one. Awash in choices, the patient could not infer the "right" eating experience. Inductive reasoning, then, may be the high-level version of a cognitive pattern evolved over millions of years, enabling animals to make the best decisions in the face of new experiences. And the reward for making the right inference, now perhaps divorced from life-or-death outcomes, may still be the stab of pleasure we get when we "make sense" of the disparate facts, ideas, and conjectures in a problem.

Ha-Ha

Goel wanted to explore how we think inferentially during mental set-shifting—where a series of thoughts focuses our attention in one direction, but to respond to unexpected information we must shift attention abruptly. And he wanted to have some fun with it. Jokes generally present us with a setup line—*A priest and a rabbi . . .*—that points to one interpretation. Then the punch line produces an unexpected, often taboo, juxtaposition leading to a novel conclusion. With this zinger, the mental set-shift yields the reward of amusement. We laugh. So what happens neurophysiologically— from the setup line to the surprise that occurs at the cognitive "I get it" and then laughter?

Human lesion studies suggested that right-hemisphere patients appreciate humor less than those with left frontal lobe injury. So that was what Goel and Dolan expected to see when in 2001 they scanned volunteers as they listened to two kinds of gags. The people were subjected to thirty semantic jokes that played on word meanings: *Why don't sharks bite lawyers? . . . Professional courtesy.* And thirty puns that played on word sounds: *Why did the golfer wear two pairs of pants? He got a hole in one. Why was the computer tired? It had a hard drive.* (Goel's then ten-year-old son supplied most of the puns.) As a baseline condition, the scientists pitched the same setup lines, then offered up bland punch lines. In the golfer joke, the baseline was: *It was a very cold day.*[15]

They were surprised to observe that puns were routed through different neural processing trackways than semantic jokes. The puns, which involved logic and language processing, displayed only left hemisphere activity. The puns—*What kind of lights did Noah use on the ark? Flood lighting.* (Goel laughs. He thinks this is funny.)— turned on the left frontal region, Broca's area, associated with speech, and the insula, involved in assessing the qualities of taste, especially distaste, and other sensory quality assessments. The bland punch lines did not tickle these areas to the same extent.

Unlike puns, however, semantic jokes—*What do you give the man who has everything? Antibiotics*—engaged both temporal lobes. This was surprising, because language processing is generally confined to the left temporal lobe. But the right temporal lobe can also be involved in the processing of unusual word meanings, metaphors, unconventional usage, as we will see in studies of creative thinking. Contrary to expectations, they found no common area activated in the right prefrontal lobe for either type of joke. The mental set-shift took place without the help of the right PFC. Jokes must not be that taxing, not like the heavy lifting in puzzling over some ludicrous syllogism.

When they looked separately at cognitive and emotional pathways, however, they discovered that brain operations correspond to what all comedians know: getting a joke isn't the same as being cracked up by it. When a participant laughed, out-loud-rolling-on-the-floor, at a gag, a specific region lit up regardless of whether it was a pun or a semantic joke. If the person was not amused, the region stayed quiet. This special area was the ventromedial prefrontal cortex. Besides everything else it does, the ventromedial PFC, it seems, is the brain's comedy central. Like its neighbor the orbitofrontal cortex, the ventromedial is involved in reward-related behaviors. In enjoying a gag, the brain is in effect rewarding itself for being clever, Goel surmises.

What's the advantage of having a mental joke meter? When we think beyond the obvious, Goel suggests, we're using more sophisticated mental operations. And the ventromedial PFC's little laugh bonus is evolution's way of encouraging us to do some light cognitive calisthenics. Dolan sees the findings as giving neurobiological credence to the idea that laughter is therapeutic. No one has yet looked at the specific neurochemicals that well up during the appreciation of a side-splitting gag. This study was the first imaging work to show the neural correlates of humor in healthy people. Looking at normal people is why, Goel suspects, his findings differ from previous experiments with lesion patients. What the lesion experiments may have detected in the patients' right hemispheres was the difficulty they were having doing the mental set-shifting necessary to get the jokes.

But getting a joke is not the same brain operation as being amused by it. Surprise, a necessary condition for the mirthful experience, Goel realized, often derives from the fact that the "punch" advanced by the punch line is physically impossible or socially forbidden. Children's humor, for example, typifies violations of physical reality. Goel points to Road Runner cartoons, where Wile E. Coyote routinely plummets a thousand feet over a cliff followed by an anvil and piano that inevitably land on top of him. He always picks himself up and walks away to fall another day. Here's another joke of this type: *A man goes to a psychiatrist and says, "Doc, my brother's crazy, he thinks he's a chicken." The doctor says, "Why don't you turn him in?" The guy says, "We would. But we need the eggs."* Or this David Brenner offering: *I was on the subway, sitting on a newspaper. A guy comes over and asks, "Are you reading that?" I didn't know what to say. So I said yes. I stood up, turned the page, and sat down again.*

Adult humor, on the other hand, often pushes the boundaries of social norms—sexual , religious, ethnic—for the payoff. (The movie *The Aristocrats*, a festival of scatology, celebrates this.) And then there's a "sick" category that pushes the envelope of normal relationships toward psychosis. Jack Handey is a master of these: witness *When I die, I would like to go peacefully, in my sleep, like my grandfather did. Not screaming and yelling like the passenger in his car.* While social-norm boundary-breaking is the point, norm violation to offense can put a damper on mirth. Here's one from Jon Stewart that is more scathing than funny: *I celebrated Thanksgiving in an old-fashioned way. I invited everyone in my neighborhood to my house, we had an enormous feast, and then I killed them and took their land.*

Goel and Dolan wanted to observe the neural substrates of social regulation in joke processing. They also hoped to see if experiencing jokes in a visual, cartoony mode engages different pathways than linguistic-verbal jokes. So they picked ninety-two cartoon gags that fit into four categories: funny and socially acceptable; funny and socially unacceptable; not funny but socially acceptable; and not funny and socially unacceptable. Sixteen volunteers rated this array of jokes while in the scanner. From the firing patterns it was clear that for cartoon jokes just as for verbal ones, the experience of getting the joke—experiencing the norm violation—excited an orbitofrontal PFC-amygdala network, while finding it funny fired up the ventromedial PFC comedy central system.

In the course of the experiment, however, Goel and Dolan discovered their volunteers had fallen into two groups—those who were offended by many inappropriate jokes, and those who enjoyed them. The enjoyers were open to jokes of all stripe, and offensiveness was not a factor in whether they were funny or not. But for the eight who found some gags offensive or socially inappropriate, their irritation corresponded to a loss of funniness. Most brain-firing patterns were common to both groups. When both experienced social-norm violation, a left orbital PFC system lit up. The left orbital PFC has been thoroughly implicated in social processing and the perception of social cues. And people with damage to these regions can be socially handicapped. As we'll discuss, the amygdala, the OFC, and temporal areas constitute a system some call the "social brain." Thus, firing in the experience of norm violation in jokes supports this network's role as part of a social brain system. And when a joke was funny for everyone, the ventromedial PFC comedy central network fired as in the previous study.

But the firing patterns of the two groups varied at a crucial point. When the turned-off subjects grew less and less amused, there was increasing activity in the right hippocampus, even as the ventromedial PFC was turning off. The no-joke-too-nasty group had vigorous firing in the funny-bone ventromedial PFC system, including the nucleus accumbens, while their right hippocampus was relatively deactivated. The offended participants' response to the norm-violating, off-color jokes, then, crossed some acceptability threshold, and weakened their experience of mirth. Their neural response to increasing offense was expressed by increasing firing in the left orbital PFC, as with the other people. But their weakened funniness rating in the face of norm violations resulted in relative deactivation in the ventromedial PFC.

So try it out. Here are a few jokes of varying incivility. How you are or aren't amused by each is being played out in the dual joke meters of the PFC.

> From Jack Handey: I think a good gift for the president would be a chocolate revolver. And since he is so busy, you'd probably have to run up to him real quick and give it to him.

> From Chris Rock: The only thing I know about Africa is that it's far, far away. About a thirty-five-hour flight. The boat ride's so long, there are still slaves on their way here.

> And one from Goel's repertoire: What's yellow and green, stinks, and lies on the side of the road? A dead Girl Scout.

IQ and the PFC

"It taxes your brain when you simultaneously work on one problem, while keeping another piece of information tucked away, but ready to go," says Stanford's John Gabrieli, "when you are doing both 'this and that,' and mustn't get confused on any part of it or else the whole thing falls apart. Many subordinated tasks are involved in this capacity to orchestrate several things in your head at once. And how a person succeeds in this orchestration is predictive of how he'll do on a variety of intelligence tests."

We have a love-hate relationship with intelligence tests. We fear them and consider their conclusions to have little bearing on our status in life. We judge them cruelly unfair, culture-bound, and too narrow in scope to measure the cornucopia of mental capabilities that humans display. And yet we make a white-hot fetish of IQ and IQ tests. A Google search of "IQ Test" dredges up a million hits in countless permutations and many languages. Intelligence, in fact, is a huge factor in our perceived status in life.

But is there a brain circuit congruent with "being smart"? With ever more studies revealing the PFC as master agent of thought, it was only a matter of time before an attempt to correlate prefrontal activity with intelligence was consummated in the lab. Leading the charge was Adrian Owen's colleague, the Cambridge neuroscientist John Duncan. Duncan's team boldly titled their 2000 *Science* paper "A Neural Basis for General Intelligence." Predictably, the popular press pounced on it, with headlines blaring: the brain's "intelligence zone" found at last! Of course it was a bit more complicated.

Indeed, to appreciate what Duncan hoped to identify, a few words about the idea of general intelligence are in order. By the early twentieth century, tests of mental ability had proliferated like weeds: tests of visual and spatial abilities, abstract reasoning, special competencies in reading, math, memory, and so on. Cutting a swath through the data and correlations, the British psychologist Charles Spearman in 1904 concluded that a single common factor accounted for across-the-board high scoring. He dubbed it the "g factor," for general intelligence. A "high-g" person scored in the superior range on a diverse array of tests, because he or she had a specific brain region finely tuned to generalized problem-solving. A "low-g" individual did poorly across the same psychometric measurements due to subpar performance from this yet-to-be-determined seat of intellect. That most IQ tests involved some degree of abstract reasoning led Spearman to posit "g" as the true and maybe genetically determined essence of intelligence.[16] And g has held up well over time. Standard IQ tests and g correlate with success in a range of lab and real-life situations, suggesting there might be common processing in much complex mental activity.[17]

As Spearman was promulgating g as a basic property of intelligence located somewhere in the brain, another British psychologist, Godfrey Thomson, offered the opposing view: that various kinds of intelligent thinking require a "diffuse recruitment" of skills, drawing upon the contributions of many neural information processors. So, Thomson believed, there could be no single g factor but rather multiple forms of intelligence. Debate between these camps has sparked a kind of Hundred Years' War of the Mind. Among other things, Duncan wanted to know who was right, Spearman or Thomson. Is there a detectable neural basis of intelligence? If so, is it a relatively confined set of structure/functions, or diffusely arrayed? Duncan admitted that he was biased toward the g factor, and from years of studying the PFC suspected he might find the locus of this intelligence processor somewhere in the frontal lobes.

Using a PET scanner, Duncan and colleagues peered into the brains of sixty volunteers, ages twenty-nine to fifty-one, while they sweated over a range of problems. (Having begun this study before fMRI technology was available, Duncan decided to follow through with the older PET device.) The scientists used brain-twisting, high-g verbal and spatial problems culled from a standard collection of putatively nonbiased g intelligence tests called "Cattell's Culture Fair." A sample verbal question, for instance, asked participants to identify which of the following sets of letters differed from the others: LHEC DFIM TQNK HJMQ.[18] The team developed easy, low-g problems as controls. In the low-g version of the verbal question, though, the participant merely had to find the one set in the sequence OPQS GHIS LMNO IJKL, whose letters were not in strict alphabetical order. The spatial tasks were likewise calibrated as either high- or low-g.

The results confirmed Duncan's pro-Spearman bias. High-g tasks did not specifically recruit multiple brain sectors. Instead, one region lit up like Broadway. "The data strongly favor the hypothesis that lateral frontal functions are selectively recruited by high-g tasks," Duncan wrote in the *Science* paper. The team saw that the left lateral PFC fires more intensely for verbal tasks, while spatial problems invoke lateral prefrontal areas in both hemispheres. There was also some activity in the anterior cingulate cortex and visual areas in both the occipital and parietal lobes. Nonetheless, the investigators were surprised to find that brain activity was *that* specifically localized to lateral PFC areas for both problem-solving domains. Duncan went so far as to assert in the *New York Times*, "What we're seeing here seems to be a global workspace for organizing and coordinating information and carrying it back to other parts of the brain as needed."[19]

Although Duncan's team did not rate individual differences in the PFC activity of their subjects, the take-home message was inevitable: the better your lateral PFC does its job, the higher your g. "Some people are blessed with a workspace that functions very, very well," Duncan acknowledged.[20] Inevitably, opponents of g theory leapt to refute this conclusion as well. In *Science*, the Yale psychologist and intelligence expert Robert Sternberg wrote a critique titled: "Cognition: The Holey Grail of General Intelligence," noting that g may not be a good measure of intelligence, since g tests do not measure talents such as creativity or adaptability.[21] In some experiments, mentally more adept people, he noted, sometimes show less PFC activity during analytical tests than lower-IQ scorers, suggesting that Mensa types need not work as hard as dimmer bulbs.

This is a good point, but it may reflect more on the tasks lab participants are subjected to than anything else. It may well be, as Duncan remarked, that "people who are having trouble are spinning their wheels to solve the problem." Anticipating the outcry, Duncan avowed in the *Science* paper that the lateral PFC is not the *only* active g-spot, but that "g reflects the function of a specific neural system, including as one major part a specific region of the lateral prefrontal cortex."

Adrian Owen admits the whole intelligence dispute is a sociopolitical hot potato, and concerning the critics' howls, in his nononsense way he voices the obvious. "I'm not sure why they have a problem with this. Some people have visual cortices that don't see so well either: people with motor cortex problems can't move their arms around." Given that the PFC is dedicated to thinking, it only makes sense it would work better for harder questions, and work better in some people than others. In a later discussion about g, two eminent psychologists, Britain's Robert Plomin and Stephen Kosslyn at Harvard, added, "Not all researchers are comfortable with the idea that a single factor may influence all types of intelligence." And although g may not be the whole story of intelligence, "trying to tell the story without 'g' loses the plot entirely."[22]

In 2003, a team led by Yale's Jeremy Gray, then at Washington University, and Todd Braver leapt into the fray by conducting the first large-scale fMRI study to probe individual differences in g intelligence and PFC functioning. They asked: Are individual differences in reasoning ability reflected in higher or lower activity in the lateral PFC network? Do scores on intelligence tests mirror important frontal operations? And if so, why?[23]

As did Duncan, the Gray-Braver team started with the premise that g is the most valid, culture-free indicator of mental ability. But by now g had evolved into "gF," the F standing for "fluid"—that aspect of intelligence less tied to any one specific ability or "crystallized intelligence," such as language or mathematical skills. One's aptitude in a gF IQ test is thought to reflect one's ability in spatial, memory, perception, verbal, and numerical tests. Duncan thinks the fluid g is a good measure of your ability to come to grips with a multiplicity of problems, and that each person's dorsolateral PFC is endowed with more or less talent for this problem-solving. What is this part of the brain doing in some people that it isn't doing in others? Weighted more heavily toward spatial testing, gF is supposed to further reduce the scoring disparities among ethnic or socioeconomic groups, and to stress more "pure" abstract reasoning and novel problem-solving acumen. Braver and Gray, moreover, proposed that gF may be related to metacognition—more abstract thinking. This may be what Goel saw in the judgment of an inference's "rightness" in inductive reasoning.

A classic measure of gF is the Raven's Progressive Matrices tests. The test was developed by Spearman's student John Carlyle Raven, who, perhaps tellingly, was profoundy dyslexic and suffered grievously in school. A Raven's matrix generally consists of sequences of abstract visual pattern sets, usually arrayed within a three-by-three-square matrix—nine squares—like a tic-tac-toe box. Generally the lower right-hand square is empty. The task is to select from a multiple choice of supplied patterns the correct one for this empty box.

In an easy version, one might see a line of three circles, each larger than the next, followed by a line of squares, each larger than the next, followed by an incomplete line of triangles—the third box is empty. But from a selection of several shapes one quickly chooses the triangle equivalent in size to the biggest square and circle. This is a one-relational Raven's matrix (see figure 3 on page xi).

After that the test grows harder, with two or more relations. To fill in the empty square, you have to discern a rule for what determines the vertical pattern, as well as a rule for the horizontal, and perhaps the diagonal pattern. Trying to infer these ever more multiple and abstract spatial principles for why you should choose one doodle-like shape over a pretzel shape to complete a matrix can bring on acute prefrontal pain—and make you quoth, "Nevermore!"[24]

Gray and Braver gave forty-eight volunteers a brain-cracking Raven's Advanced Progressive Matrices test to rate each participant's intelligence status according to high or low gF. Then they scanned the subjects while they performed a challenging working-memory test, called the "3-back." Here the participants viewed a series of words or pictures of faces every 2.3 seconds, and were asked to indicate whether the word or the face was identical to the one that appeared on the computer three images before, a test that required constant updating and evaluation of the contents of working memory. To make matters more difficult and confusing, sometimes the series were interspersed with "lures"—faces or words that had popped up two, four, or five images before. When lures appeared, the test-taker's accuracy plunged and response time increased.

Nonetheless, high-gF IQ participants not only performed with more accuracy but also displayed more fiery brain activity in the scanner than the low-gF scorers on the tricky lure trials. And what neural regions lit up more brightly in the higher-scoring subjects? Primarily the lateral PFC, the cerebellum, and, predictably, the anterior cingulate cortex. So, in the mechanics of mind, not only is the PFC more strongly recruited by those endowed with better general intelligence, so also is its cognitive enforcer, the ACC. High general intelligence seems to involve both better PFC function and elevated ACC activity that "tells" the PFC to pay attention, tighten focus, in the face of distraction. The more active, high-gF network strengthens a person's mental cool under fire, mental toughness at crunch time.[25]

IQ and the Cocktail Party

The Braver-Gray experiment drew on a theory of intelligence proposed by the Georgia Tech psychologist Randall Engle. According to Engle, the ability to block out or overcome mental interference is a cardinal feature of high-gF intelligence. Better concentration, better capacity to ignore distractions, superior alertness to conflicting information streams—all amplify mental performance; indeed, cognitive control might *be* the defining element in high IQ. Arguing that working-memory capacity is responsible for these control skills, Engle and colleagues posit that working memory and g are highly correlated, if not identical.

A former student, Andrew Conway at Princeton, recently tested Engle's theory by means of the "cocktail party phenomenon." In the mid-twentieth century, psychologists investigating attention spans discovered that around a third of all people detect their own name being spoken when they are deep in the midst of conversation. Some people who are involved in a noisy environment find their attention captured by the sound of their own name—and others do not. Conway and colleagues used this "cocktail party effect" to see if people who have difficulty screening out distractions are those who also show relatively lower working-memory capacities, thus lower IQs.[26]

The investigators tested the working-memory abilities of forty undergraduates, categorizing participants according to high and low working-memory span. Next, each subject listened to and repeated a recorded message played to one ear and attempted to ignore a different message simultaneously presented to the other ear. Unbeknownst to the subject, his or her name was also presented in the unattended message. The difference was fairly startling: only 20 percent of the high-capacity working-memory performers reported hearing their names, while 65 percent of the low performers did. The low working-memory subjects also made more errors on the repetition task. "Yeah, that was a really neat outcome," Conway says. "Everything we'd been studying suggested that low-capacity people have problems with selective attention: maintaining what they're doing in the face of interference, in screening out distractions."

The results have telling implications in a world where we are bombarded by entities vying for our attention and messages that attempt to frame our reality with knee-jerk reactions. Many people come up to him, Conway says, with anecdotal stories. "Like they have a spouse who is a very smart professor, and when he's focusing on his work you have to basically hit him over the head with a two-by-four to get him to look away." Or others who are so easily distracted that all you have to do is walk by them to cause them to lose their train of thought. "We sometimes use it as a joke," he confesses. "If we're in a lecture and somebody opens the door in the back of the hall and walks in, it's fun to watch who looks back and who remains intent on the talk." Conway's grad students are doing a study in which they videotape an introductory psychology class while they do distracting things like drop pens on the floor and see who looks around.

"I look at this high selective attention/working-memory capacity as efficiency," he continues. "When you are performing well, you're maintaining activation to brain areas that contribute to doing the task, and keeping activation away from areas that represent irrelevant information. I argue that working-memory capacity as a function of this control drives my ability to maintain a goal and thus drives performance." Studies of self-control in children support Conway and Engle's ideas. The Florida State University psychologist Roy Baumeister, for instance, has shown that preschool children's degree of self-control was predictive of how well they would later score on standardized cognitive tests.

Baumeister did some rather devilish things such as putting a four-year-old in a room with a dish of M&M's, then telling the child as he was leaving the room, "If you don't eat any candy while I'm away, you'll get a whole bag later. But if you eat any of these, that's all you'll get."

Videotapes show kids staring at the M&M's, putting them up to their mouths, putting them back. Some children broke down and ate them. But how well they held out for the later reward did predict their cognitive performance in years to come.

Many parents ask Conway if there is anything they can do to improve their children's cognitive control capacity. Not a lot, he replies. "I do tell them to encourage their kids to read. Reading not only increases verbal ability, but it increases ability to sustain and control attention. Reading is attention-demanding: you must focus on what otherwise are arbitrary scribbles on a page. And you have to update your mental representation of what's happening in the text and block out distractions to do that. It's great that children are reading eight-hundred-page Harry Potter books. Everybody should read more, regardless of what it is."

Conway will only speculate on the attentional merits of video games and television but suspects neither is worth much. "Whereas reading requires what I call 'endogenous attentive control'—you have to control where you go—TV and video games are the opposite. Control is placed on the screen and all you do is react. That 'exogenous control' could have a negative impact. But I don't know if anyone has rigorously tested that," he adds, not wanting to say that video games are bad. "I'd be getting all sorts of phone calls from Sony." The rebuttal to people who claim their kids are focused when they play video games, he asserts, is, "They're not focused! They're just following blips. The screen is what's guiding them. The game is guiding them; they're not guiding the game." In reality, gaming programmers call the shots, and the games are unremitting Pavlovian exercises in conditioned response. Conway allows he does like the children's TV show *Blue's Clues*, however, where every episode is a little mystery. "The kid has to maintain in mind three or four clues throughout the half-hour show to solve the mystery, to maintain and keep updating the goal with each new clue. It's a great working-memory task."

Conway also cited work by the Stanford social psychologist Claude Steele showing the negative effect of stereotyping on intellectual performance. If you bring women into a room to do a spatial reasoning test and tell them that "we know from years of experience that men outperform women on these tasks and your work will be used to support this thesis," this will "threaten" the subjects into performing worse than they would have had you said nothing. Steele's work shows you can "stereotype-threat" just about any group, not just minorities, if you find the right threat. "Minorities are probably put into these situations more often than nonminorities," says Conway, "but everyone has his hot-button issue. If you're put in a situation where your issue becomes salient, it's going to have a negative impact in your cognitive control." Working memory will be unable to block out the distraction of the stereotype threat; negative emotion crawls beneath the barriers set up by cognitive control and affects performance.

There is evidence, as well, that g intelligence correlates with the ability to suppress undesirable thoughts, such as racist images. When subjects in a lab test view an easily stereotyped individual— say, a bearded man wearing a turban—they show increased stereo- typing behavior, despite their efforts to the contrary. They are, as the jargon puts it, under an extra "cognitive load." And while their PFCs are engaged in other tasks, they may not be able to screen out stereotyping's fixed thought structure that pops into their minds, despite their best intentions to inhibit it. The ability to suppress such thoughts and mitigate their impact varies across individuals and may correlate with working memory/general intelligence.

Blaming the Victim

Fluid g and working-memory capacity may predict not just elite, ivory-tower smarts, success in the marketplace, or high income but also how you parse the nuances of everyday reality. Similar to stereo- typical thinking, "counterfactual thinking" is another cognitive phenomenon that is affected by working-memory/intelligence effi- ciencies. Counterfactual thinking is the common yet fairly complex mental act of wondering about "what might have been." It is the thought mode that gives rise to fantasies about big historical events such as: What if the Nazis had won World War II? Or what if John Kennedy had not been assassinated? Or Al Gore had been in the White House? as well as more quotidian ruminations such as, What if I'd been at that party he had attended? The "road not taken" is a venerable theme in novels and movies, notes the Arizona State Uni- versity psychologist Stephen Goldinger, who has studied the conflu- ence of counterfactual thinking, working memory, and IQ.

Counterfactual thinking is something we do readily, designing representations of alternate realities—like so many quick-acting computer sims with hypothetical inputs and novel outcomes—to events that have occurred. People spontaneously generate counter- factuals in the face of bad events. (Had a certain distinguished neu- roscientist not jaywalked across a busy street in Connecticut at the exact moment she did . . .) A standard example is, If only I'd shut the window! after one's house is burgled. Or the "what if" thinking one indulges in when arriving at the airport gate as the plane pulls away.

Unlike stereotypical thinking, counterfactual thinking may have an heuristic "coping value" in conjuring ways you might have done something better to avoid the misfortune that befell you. Counterfactuals are a way to think through outcomes without putting yourself at risk. They are an adaptive form of planning as you play mind games: What if I tell M. she should get help for her drinking? What if I don't? Counterfactuals are a way of maximizing your likelihood of success.

On the other hand, counterfactual thinking can have bad side effects. Counterfactuals are often distressful, for one thing. Losing in a near win is painful. Think of an Olympic silver medalist who ruminates about how he might have run one step faster after a one-hundredth-of-a-second loss in a five-kilometer race. Or the pain losers experience in their postmortems after a closely run political race ("If only we'd called him a lying bastard"). Imagine, offers Goldinger, that your standard set of lucky lottery numbers finally won, but you skipped buying a ticket that week. "Something like that that happened to me," he confesses, laughing. Although a counterfactual mourning period for the near miss is natural, it also poses the risk of an unhealthy obsession.

Counterfactual thinking, furthermore, can co-opt logic, warping perceptions of causality into inaccurate assessments. To evaluate the truth in a situation, you often must suppress or discount the natural upwelling of counterfactual thoughts. Goldinger wanted to know if general intelligence correlated with a person's ability to squelch counterfactuals' biasing effect on reasoning, especially when he or she was already working "under the cognitive load" of processing multiple ideas in a difficult problem. Under the working-memory stress of a cognitive challenge, would lower-intelligence people "take the counterfactual's mental shortcut and just go with it?"

Civil court is a context in which counterfactual thinking comes into play with negative consequences, possibly affecting a jury's judgment and thus the outcome of the case. So to test counterfactual thinking and intelligence, Goldinger and colleagues used scenarios of civil trials wherein the "jury" must decide how much compensation to award the victim in an accident case.[27]

Among juries, there is a counterfactual tendency to blame the victim. Say Paul customarily leaves the office at five thirty and drives home. One evening on his daily route, he is broadsided and badly hurt by a driver who ran a stop sign. Hearing this story, most juries would recommend compensation for Paul and perhaps punishment for the other driver. But, poses Goldinger, what if Paul had left work earlier and gone to the movies, and on his way to the cineplex the accident happened? Although the accidents are identical, in the second scenario people might note Paul's "cavalier behavior" and decide that lower compensation is due.

The thought "If only Paul had stayed at work until five thirty" might come to the juror's mind. This counterfactual should have no bearing on the rational assessment of compensation. But these thoughts do sneak into our reasoning and somehow "get to us." The opposite result is also possible. What if, instead of shirking off work to see the latest Austin Powers flick, Paul got an emergency call to return home and was rushing to deal with a domestic crisis when he collided with the stop-sign runner? The accident now appears "tragic," and compensation judgment may thus be larger. Counterfactuals, then, are not unlike other automatic biasing thought processes, says Goldinger, "wherein ideas unwittingly spring to mind" that require suppression.

In "Blame the Victim," Goldinger first tested 138 students to rate their working-memory skills/general intelligence and picked the highest- and lowest-scoring students for the study. Then he elicited the student "jury's" judgments on virtual court cases—both straightforward cases and those containing counterfactual elements. In some of the tests, however, the students had to make their judgments while they were simultaneously preoccupied with keeping nonsense information in mind—a working-memory task that siphoned mental energy and interfered with their deliberations. How much would the students' ability to shut down their counterfactual thinking be affected by the extra mental "noise" piled onto their PFCs? What was the effect of "loading" working memories onto the executive processes as they also weeded out irrational counterfactuals? Goldinger suspected people with higher working-memory capacity would better control counterfactual inference. In this sense, "Blame the Victim" is a more complicated version of Andy Conway's "cocktail party phenomenon" study.

As in "Paul's Accident," there were two versions of each courtroom drama. In one, the victim behaved habitually before the accident; in the other, he or she did something "different" just before disaster struck. Although the atypical acts had no causal bearing on the victims' rendezvous with fate, these detours served as "bait" for inducing counterfactual thoughts, tempting the student-jurors into rendering a less compassionate verdict. One vignette, for example, involved Mark, a basketball season ticket holder. In the control version, Mark attends a game, sitting in his usual seat. A light fixture falls from the ceiling, breaking his foot. In the counterfactual version, Mark takes advantage of an empty seat closer to the floor of the arena, and the light falls on his foot.

The students' primary task was to render a verdict on the monetary compensation to the victim. In Mark's story, for instance, the students had to decide on a financial settlement with the arena's insurance company, ranging from $5,000 to $95,000. When Goldinger compared the judgments of the two groups, he found that students with lower working-memory capacity were more susceptible to biases of counterfactual thinking (blaming the victim) than the higher-octane working-memory group. The lower working-memory group awarded Mark less money in the counterfactual scenario in which he moved to a different seat. But significantly, these lower-capacity students succumbed to counterfactual distortions only when they were also simultaneously preoccupied with keeping in mind the irrelevant working-memory task. This suggests that counterfactual thoughts, arising automatically in all of us, may require effortful, demanding prefrontal activity, a kind of suppression that "more intelligent" people do better. No one has found any correlations between intellect and generating counterfactual thought—only the ability to suppress it when necessary.

"Only when people were making judgments did the effect of the counterfactuals come out. Everyone was perfectly capable of getting the information set up in a mental scheme to understand the story— even when they were holding these nonsense syllables in memory. But when they were asked to make a judgment, the higher-span people were able either to say, 'That's not relevant. I shouldn't let it affect my judgment,' or to not have the counterfactual thought distract them at all," says Goldinger, who admits his group doesn't know what's going on in the evaluative processes in the lower-span people, although "the counterfactual is clearly getting to them. Maybe it popped into their heads, they're not discounting it but letting it affect their judgment, or maybe they're trying to set the notion aside, but it's still leaking out into judgment." Goldinger is pursuing ways to dynamically track the subjects' thought processes throughout the sequences of events. "It is so interesting the way this prefrontal cognitive control," he muses, "is so susceptible to degeneration."

Testing Your Fluid G

Is superior working-memory processing in the PFC the sole determinant of intelligence? "Clearly there's gotta be different dimensions of intelligence," Todd Braver offers. "I guess I'm kind of conservative, but I feel intelligence is so multifaceted, if you measure intelligence with tests, it may be different than if success in life is the yardstick. Who knows? I don't want to place my bet on one thing."

The British seem even more focused on intelligence quotients than North Americans, and when the flurry of imaging studies on the brain bases of IQ appeared, the London *Times* offered online readers the occasion to test their fluid g with mind exercises. Here are a few:

- Remember and dial a ten-digit telephone number while people are having an interesting conversation around you.
- Picture the map of directions to your destination while having a conversation with the passengers in your car.
- Ask someone to read the words "dog," "cat," "chair," "table," and so on, to you in random order, one word every two seconds, and try to tell whether the word you hear is the same as the one you heard three words previously.

Some other neurobic exercises include:

- Use your nondominant hand when you do your morning rituals such as combing your hair, brushing your teeth, or making breakfast.
- Turn the pictures on your desk or shelves upside down.
- Immerse yourself in unfamiliar surroundings, if possible where the foods are new and no one speaks your language.

The eminent researcher of cortical development in early childhood, Adele Diamond, thinks that bilingualism is a good way to enhance a child's working memory and control processes. "Because bilingualism puts a heavy demand on inhibition and selective attention, we think being bilingual pushes precocious maturation of the prefrontal cortex," says Diamond, the founder and director of the Center for Developmental Cognitive Neuroscience at the University of Massachusetts Medical School. "These are skills needed later at school—to be able to filter out distractions, pay attention to a teacher in a noisy room, or stay at a task without being distracted by this, that, or another thing. Being able to stay on your seat and be obedient—all these put heavy demands on the prefrontal cortex. If you are a little more developed, you can get a significant leg up."

To parents wanting to maximize their child's prefrontal development, Diamond offers commonsense advice: give your child activities that tax inhibition and require selective attention, such as learning languages, playing sports or a musical instrument. "You want to encourage the child's natural curiosity, to learn that exercising the mind is fun. That will take children much further than any specific thing you teach them. But it's a terrible idea to push a child; it will be more detrimental than if you did nothing."

Metamind: On the Inside, Looking In

What role does introspection—reflecting on what you're thinking—play in rational thought? And how does the PFC package these programs into something like what the AI guru Marvin Minsky calls the "cluster" or "suitcase of consciousness"?

Imagine that a person on a long hike in a great northern wilderness finds himself off the trail and irrevocably lost. The hiker will create a primary goal, a big picture of finding his way back to the trailhead. And unless he panics and his prefrontal cortex shuts down in adrenaline-fueled chaos, he will begin modeling in his mind sets of subgoals: strategies for keeping warm; conserving food, water, and physical energy; orienting himself with map and compass bearings; analyzing features of topography, streams, ridgelines, and vegetation. (Of course, if he has a cell phone and calls for help, that's another story.) The adaptive advantage of this organization is powerful—a cognitive processing schema that is "layered" from most immediate (staying warm) to most abstract (the representation of his car parked far away, and then further abstracting out to the essence of "safety").

In advanced reasoning, a sector of the PFC computes more abstract, big picture processes. The ability to focus on goals and suppress competing thought processes is not entirely dependent on the lateral parts of the PFC. Complex mental workouts require yet another layer, a superstructure in the mental representation you've built, a space where you can say, "If I did A, then what will happen after that?" and hold that online while comparing it to, "Suppose I did B instead? What will that result in?" In these computations, you require the frontopolar region of the PFC.

Brodmann area 10 is the apex of the human prefrontal cortex. Situated at the frontmost tip of the cortex, it is larger compared to the rest of the brain than it is in apes. Quite possibly the neural apparatus associated with area 10 is what expanded most during hominid evolution. Area 10 is known as the frontopolar cortex, and in some ways that name is fitting, summoning up the extremes of Earth's magnetic field, the frigid distances from temperate life and the fantasy world of Santa's workshop.

What are the principles that unify the frontopolar region's recondite functions? Two related theories describe what's going on up there. One posits that area 10 tackles the most abstract mental processes; that functionally as well as structurally, the frontopolar cortex (FPC) is the pinnacle in a hierarchically arranged series of PFC subsectors, stacked wedding-cake-like from ground-level zones, such as the orbitofrontal and ventromedial, on up through the ventrolateral and dorsolateral, to the poles.

Washington University's Todd Braver is testing a version of this hypothesis, espoused by the French neuroscientist Etienne Koechlin. In the late 1990s, Koechlin and colleagues proposed that the anterior PFC comes online to "help" the "lower" PFC conduct large-scale working-memory operations, such as, say, Goel's architectural design scenario, wherein you need to sustain a big-picture, long-term goal while simultaneously allocating brain power to accomplish subsets of this goal necessary for its eventual success. The frontopolar cortex thus keeps a person intent on the distant prize, while the rest of the PFC and its partners explore options and alternatives for getting there. In a newlyweds' house-buying scenario, for instance, the couple explore various mortgage rates, while keeping in mind the price of the desired house. Or a quarterback holds in mind his goal of making a game-winning touchdown within the last three minutes of the fourth quarter, while driving down the field with a series of plays designed to exploit specific defensive weaknesses and simultaneously burn the clock.

Braver further proposes that the anterior PFC is engaged in pre-serving this big picture (the hiker's car at the trailhead, the quarter-back's touchdown, the couple's new house), this mental video clip that needs "protection" from "noise" arising from subgoal process-ing details or unproductive emotions. The frontopolar cortex inte-grates subgoal computations with the overarching representation, combining and updating the two data streams into one confluence, the grand goal. The quarterback integrates new information he receives from what happened on the last play into the overall offen-sive strategy and so modifies his plan to better drive to the end zone.

To test the idea, Braver needed to isolate frontopolar activity from other PFC activity. He designed a stripped-down semantic test with simple goals and subgoals. He asked scanned subjects to keep sets of words in mind while they simultaneously classified them in abstract categories. The results confirmed Braver's suspicions: the frontopolar region uses a special representational code to integrate and hold online the goal, while the rest of the PFC is working on implementing subgoal processes.[28]

The steadiness and durability of the higher, long-term represen-tations in area 10 may enable you to more deftly coordinate lesser, rapidly changing details. Braver's theory implies a temporal hierar-chy rating the "average lifetime that a representation or a system will hold on to information." This gradient of duration implies that the lower, ventral PFC regions carry and hold more transient data, information we don't need to hang on to for a long period. The dorsolateral PFC, then, distills information "welling up" from the ventrolateral PFC and may need to maintain it for longer periods of time. The frontopolar territories, wherein the information is more abstracted from the outside world, require its representations to last for longer periods of time.

This hierarchical thinking may be what separates human cogni-tion from other animals' thinking-in-time, a unique genius for reaching a goal through a series of steps enacted (perhaps linearly, perhaps not) toward a vision of the future. Humans are stairway builders, and if one imagines buildings as external metaphorical instantiations of our cognitive architecture in some ways, one can see this prefrontal architecture as existing on multiple levels with rooms of many sizes.

With his colleague Randall O'Reilly at the University of Colorado, another in Jon Cohen's coterie of neural-net aces, Braver is implementing a connectionist computer model of hierarchical, time-scaled, increasingly abstract information-processing circuits. This next iteration of the PFC "cyborg" has a "frontopolar" structure added on to its units. As we will detail in a later chapter, this evolved silicon model uses a series of "outer loops," where information is maintained over a long period of time, while shorter-term information is updated within its "inner loops." The outer loops enable the neural net to contextualize what it needs to represent in the inner loops. "We're saying to it, 'Here's the kind of task the system needs to do,' and let it run," says Braver. "We're starting to see evidence that these inner and outer dimensions indeed emerge under the pressure of learning how to do complex tasks. The model spontaneously self-organizes to best handle this task."

While Braver's team posits a hierarchy of abstract thinking units, reaching a summit in the frontopolar cortex, and attempts to build a machine simulating this dynamic architecture, another young scientist is developing a parallel theory for metaprocessing in the PFC's polar zones.

Metacognition

In her ambitious search for a framework that explains the richness of prefrontal functions, Kalina Christoff, like Braver, arrived at area 10. Christoff, who grew up in Sofia, Bulgaria, got her PhD in psychology in John Gabrieli's Gab Lab at Stanford. Her life was a "curvy trajectory," she admits, from economics to neuroscience. But the advent of imaging technology in the mid-1990s sealed her involvement with neuroscience. This was where the action was. "Perfect timing, yes?" She laughs.

Christoff launched her exploration of abstruse reasoning and thought processes in 2000, with Gabrieli, when she proposed that this anterior tip of the PFC serves as a "third layer" to more posterior dorsal and ventrolateral stacks of PFC functioning.[29] Whereas these PFC zones are dominant when we process externally generated information, she said, the frontopolar areas evaluate information that cannot be perceived from the external world, but is generated from the contents of the mind itself. Area 10 screens the innermost mental films scripts, the plays-within-a-play.

"When you reach a conclusion, or draw a relationship between two or more things that are not perceived together in the environment—if you figure out these relationships, the form of thinking you are performing is by definition abstract," Christoff, now with her own lab at the University of British Columbia, responds to Braver's hypothesis. "One links elements of abstraction throughout all levels of the PFC, especially the lateral regions. But what humans uniquely do is go to the next level, where you think about your own thoughts—the 'introspective thought process.'" In terms of our lost hiker's brain, area 10 would be firing intensely as he asked himself the big questions: "Okay, how do I set this compass reading?" and "Am I starting to panic? How must I control my anxiety and figure a way to get out of this mess?"

As always in neuroscience, the trick is to demonstrate how the brain is metacogitating. Christoff and colleagues designed an imaging experiment using for cognitive processing fodder a Raven's Progressive Matrices test. To escalate the levels of abstraction, Christoff's team realized they needed to present three types of matrices: so-called 0-relational problems, where all the figures in a matrix are identical and so require absolutely no relational thinking; 1-relational problems, where each line of the matrix features a different type of figure and you evaluate the relationship in either the horizontal or vertical dimension; and most difficult, 2-relational matrices, which increase geometrically in complexity—it's as if you are no longer comparing apples and peaches, but must find the key relationship between arrays of apple and peach hybrids presented in variations of French tarts and American pies, offered in different kinds of restaurants (see figure 3 on page xi).

Although Raven's matrices have been used in cognitive testing for decades, this "relational dissociation" was new. "To distinguish different subregions," says Christoff, "we manipulated the complexity of the cognitive processes." Nobody had previously deconstructed reasoning into these relational elements, or had scanned people's brains while they reasoned at ever-harder Raven's problems. What drew Christoff's immediate attention to area 10 was that with the first subject, and consistently thereafter, the moment the person began to puzzle over a 2-relational problem, his or her anterior PFC would light up immediately and powerfully. "We already knew that most reasoning activates the entire PFC,"

Christoff continues. "But it was surprising that the two-relational was so focal, rather than the huge blob of activations usually seen in PFC. The one-relational was already 'serious thinking.' But moving up a level in complexity nicely isolated the anterior region."

What summoned area 10 during 2-relational reasoning was the need to manipulate information several stages beyond what the person could easily do by just looking at the figures. To infer the correct answer, you compare and integrate information, using your "mind's eye" to move the pieces around in your mental workshop, trying out various strategies to assess the one relationship that fits the matrix pattern. You solicit feedback from your "inner voices" to help you do this.

"When you can't act solely on the information in front of you, but jump back and forth between things, to create in your mind a new piece of information," adds Gabrieli, "that's what seems to turn on the frontopolar prefrontal cortex. You can't just look at a difficult Raven's matrix and say, 'Oh, the answer is three.' Sure, the easiest ones seem almost perceptual — the row of objects are getting bigger and bigger. It doesn't seem like you're doing much thinking. But on the two-relational ones, you start to go 'whoa' and say to yourself: I need to figure out some rule that will apply. And this turns on just about all the frontal cortex—including this polar area."

Self-reflection, here, is an elite form of executive control involved in selective attention, that sine qua non of general intelligence. And it is awareness. Selective attention plus self-reflection is happening when you say to yourself, "Now, I must do this very carefully." It is conflict resolution when you note to yourself that you must choose between two compelling but competing ideas or acts. It's at work when you tell yourself you screwed up royally on something and must fix it. Metacognition is the innermost feedback loop of currents of information and memory evaluations in judgments of learning (How well do I understand that new program?) or feelings of future knowing (I think I will perform well on the GREs). Self-referential thinking requiring levels of "self-value judgment" is the domain of area 10.

These metacognitive feedback pathways optimize thinking: you "see" yourself in the future or evaluate data concerning something that happened in the past: "Why is that story important to me?" Internally generated thought processes come into play when we

must retrieve information from memory stores of previous or imaginary experiences, so-called postretrieval evaluation: "Will this past
experience help me here?" Also engaging area 10 are memory-of-
the-future prospective operations, in which you remind yourself of
a prior intention to visit the Air and Space Museum next week.

Christoff is squarely in the camp of those who believe the PFC is
"fractionated" into specialized hierarchies of function. Another
group, however, thinks the PFC is a more homogenous tissue with
less specialized sets of cognitive functions. This faction suspects that
area 10 blazes away when we engage in more and harder processing, rather than a different quality of processing. Christoff understands where some of the opposition to her hypothesis is coming
from. The PFC-as-a-homogeneous-entity works well if your
research is on monkey physiology, she says, since there is less topo-
graphical distinction in the monkey PFC than in humans'. Even
though monkeys can be heavily coached to display abstract thinking, as Earl Miller showed so dramatically, they fail at abstract tasks
humans execute without a moment's training. As a neuroimaging
scientist working with humans, Christoff has had greater opportunity to observe the distinctive cognitive signature of area 10.

At one point, Christoff decided to stop calling the anterior PFC
the "frontopolar cortex," and begin viewing it as two zones: rostrolateral and rostromedial PFC (rostral, meaning "prow" or "front").
"I noticed that only the lateral (outer) part was activated during
high-level reasoning and calculating of tasks," she recalls. Others
had suggested that complex thinking processes accompanied by
emotions activated the rostromedial (middle) sectors. Suspecting
the frontopolar region had an "emotional" zone "inside" a more
cognitive processing area, Christoff now thinks this middle part of
area 10 is recruited specifically when a person is reflecting upon his
or her emotional states: "Why am I feeling so excited? I'd better
chill before the meeting."

This makes sense anatomically. All medial PFC regions, including the orbitofrontal and ventral, link powerfully to emotional subcortical circuits, and to the more evolutionarily primitive tracts
related to olfactory and visceral processes. In Christoff's scheme,
the rostromedial zones, high up in the PFC's prow, engage in meta-
emotional processing divorced from the external world. This may
be the true world of the shrink's couch, where one thinks deeply

and passionately about one's emotional life. Of course we are emoting almost all the time, but in Christoff's scheme, it is when we are
aware of and reflecting upon it that medial area 10 lights up. People with rostromedial PFC damage often have problems with their
selfhood, not realizing, for instance, how changed they are after
injury to this part of the PFC.[30]

For Christoff, the next step is to clarify how thoughts and emotions are distilled and reconfigured in abstract introspection. "If
you want to better understand some thought or act, first you think
about it, then you think about the way you are thinking about it.
This typically includes thinking about your emotions, and your
thoughts about whether you have been, or might well be, rewarded
or punished for an act if you follow one strategy or another. In introspective thinking we do not think linearly," she adds. Instead, our
thoughts remain centered in certain domains, like rational inference, but intermingle with introspections about our emotions concerning the rational and logical progressions. "After a while things
start converging," she notes. "A conclusion starts coming to us."

Christoff speculates that hemispheric differences also exist in
the way the rostral PFC handles internally generated meta-rational
and meta-emotional information. The right rostrolateral PFC might
specialize in more episodic, memory-related evaluation, while the
left rostral PFC might tend toward the manipulation of immediate
thought processes. The asymmetry of the lateral PFC "below" this
sector is pretty well established. Remember, the left-hemisphere
hindquarters near Broca's area, BA 44, presides over verbal computations; corresponding areas in the right hemisphere are devoted to
spatial processing. "We all believe that," Christoff gently mocks. "But
the more anterior we go into the PFC, the less spatial-verbal specialization there is. I've never seen anything with verbal and spatial differences there. And counterintuitively, most memory tasks that use
verbal material activate the right, not the left, hemisphere there.
And then our Raven's task, which is purely visual-spatial, activates
the left hemisphere! It's not straightforward."

Christoff is gathering data to support her notion that the rostrolateral PFC specializes in explicit processing, that only thoughts and
emotional operations a person is aware of go on there. She ponders
the idea that the PFC has distinct subzones that serve self-aware thinking, and others that serve nonreflective, reflexive thinking. Implicit

versus explicit distinctions have rarely been done in neuroimaging. It is difficult to construct experiments along implicit/explicit lines because the scanner is an equal opportunity imager: nonaware, implicit thought processes show up as vividly as conscious ones.

On occasion, we reflect upon implicit rules we've been following and explicitly decide to break them. The deliberate consideration or reconsideration of a pattern of action, behavior, or thought is an explicit use of internally generated information and, insists Christoff, a precondition for rostrolateral activity. Conversely, not all planning and executing of actions, behaviors, and thoughts may require area 10's engagement. A series of moves in a board game or at the poker table do not in themselves necessarily lead to rostrolateral firing. Nor does a hike in the woods, per se. But when the intricacy of the game, or the dire straits one finds oneself in the dark forest, requires explicit self-referential evaluation, then the rostrolateral lights up.

"Humans are so mentally flexible because we self-reflect when we need to," says Christoff. "If you stop yourself and say, 'Well, at this moment I followed this rule, but actually I want to follow the opposite rule next time'—that reflection creates a flexibility of behavior." (One thinks of the judiciary, the area 10 of the body politic, the deliberative, reflective social instrument charged explicitly with preserving and interpreting the rules and flexibility of the rules.) "Monkeys are very able to follow abstract rules," she adds. "Take Earl Miller. He trains monkeys for months and they become very good at following abstract rules. But they cannot integrate two-relations, suggesting they cannot transform, or be aware of, abstracting rules. They are not flexible in the kinds of rules they follow. The monkey cannot explicitly say, 'I'm following this rule, and it's an okay rule, but it could be made better if I change a few things about it.'"

The notion of explicit anterior PFC processing invokes the specter of—dare we say—"consciousness." Has the "Zone of Consciousness" been found in area 10? Christoff cheerfully deflects this question. "Ah, yes," she laughs, "consciousness is a dangerous thing! The more I look, the less of a 'spot' consciousness appears to be, and more of the tip of an iceberg. It is the tip of a pyramid, and nothing without the rest of the pyramid. The anterior PFC allows for this reflectiveness at the very highest levels of abstraction for the rest of the pyramid. But it doesn't have to work all the time. If you're in a familiar environment, following abstract rules without thinking about them, then the tip of the pyramid is not necessary, and you're not going to use it." (Is this why some people prefer the familiar grounds of theme park vacations? So their area 10s can check into franchise motels and zone out?)

Does area 10 work the same in everybody? "Neuroimaging studies," Christoff offers, "allow us to see things that are consistent across subjects. That's why we talk about 'processes' rather than 'content.' The content of thoughts will be personal, yet the general principles upon which people will structure this content can be universal. So while your experience of yesterday was very different from mine, both of us might today, upon reflection, try to understand the personal significance of what happened yesterday—my yesterday versus your yesterday. Both of us build a design that will be very different in content, but may be very similar in structure. Our imaging studies show there are reliable, common patterns during self-reflection that encompass a significant proportion of the people we look at."

Spontaneous Thought: Self-Reflection's Unconscious Twin?

Christoff and Gabrieli were also fascinated by a stubborn and vexing artifact of neuroimaging studies. Neuroimagers everywhere report seeing in their subjects considerable excitement sparking around the cortex during the time these individuals are just sitting there. This is called the "resting state," when subjects are asked to "do nothing, think nothing" between tasks. It is during this resting state that spontaneous firing patterns flare up in subjects' brains like auroras in the northern sky. But what is this noodling, mind-wandering, "something like living" flow of inner mental events that the imaging machines reveal streaming when we are disengaged from effortful thinking? Does this spontaneous thought depend on executive PFC processes?

To isolate it, Christoff's team turned fMRI protocols upside down and focused on the "rest" state as the condition of interest. They designed a task requiring an infinitesimally simple cognitive demand: to discriminate between a left- or right-pointing arrow with one's finger, to which they compared the resting state, since it involved no cognitive demands. While the subjects did the arrow task, the motor cortex was the predominant area of activity, reflecting the subjects' use of their finger muscles. During the resting state, however, a network of cortical regions flared up boisterously: temporal lobe structures were vividly alive, as were visual cortices and the rostrolateral prefrontal cortex. The robustness of activity was comparable to that during many highly demanding cognitive tasks. But the scientists saw relatively little activity in the other cognitive PFC areas. This temporal lobe activity suggested that long-term memory processes form the core of this spontaneous thought.[31]

This new insight forced Christoff to reconsider current definitions of how we live in our minds, including the authority of conscious, controlled executive processes. "This is the flip side of effortful thought," she muses. "Spontaneous thought is still thinking, but uncontrolled conclusions can emerge. Most controls on our effortful thinking are the result of the prefrontal cortex reaching outward, influencing and imposing structure on the rest of the brain." Uncontrolled thinking seems the opposite: the rest of the brain imposing a different sort of design on the PFC. Spontaneously occurring conclusions, then, may be based on prior experiences, wherein the temporal cortical memory structures reach out and "seduce" the anterior PFC with their powerful personal content. And something "suddenly occurs" to you. "I assume the awareness of this sudden thought happens in the anterior PFC," Christoff continues. "Yet if something is generated outside the PFC, and arises in the PFC without being constructed there, we're only aware of its survival, not its generation. In that sense there is probably a huge amount happening outside of our awareness."

In this fertile stream of thought, spontaneous retrieval of images, events, or acts may generate fractal-like swirls of thinking patterns, especially in the absence of deliberate, goal-related mental work. Long-term memory input into these thought rivers may help account for puzzling aspects of "minding"—such as the tendency to "drift," or change course abruptly. The untrammeled nature of this neural phenomenon may encourage "leaps of thought," or, more mundanely, the intrusion of certain thoughts into consciousness— the earworm or pop tune you can't drive out of your head—despite our efforts to turn them off. We may have to accept the notion of "thought permanence," Christoff adds, the idea that thought continues to exist even when we are not observing it directly.

Christoff suspects spontaneous thoughts form the armature of our individual sense of reality, "our worldview, for want of a better word." Anything that makes us feel we perceive the order and sequence of what happens in life contributes to a subjective wholeness of oneself in the world. The ongoing river of spontaneous thought enables us to spin a coherent narrative whereby events and ideas isolated in time are folded into some bigger codification of reality.

To maintain one's own weltanschauung probably requires this prefrontal circuitry to be intact. Christoff is reminded of the confabulating prefrontal patients documented by the University of Toronto's Morris Moskovitch. If an observer asked a patient to talk about his past and what kind of life he's had, he might reply that he has a wife and three children and worked as an engineer. Except he would be lying. "Not only would he be lying, but when confronted with the fact he's never been married and has no children, he would start giving reasons justifying his previous statements," Christoff adds.

"A brain-damaged patient who confabulates perhaps cannot form the narrative structure necessary to understand his own past and present. Therefore all the elements—true and imaginary—have equal standing in his mind." And he is compelled to construct his life-movie out of these fragments. Since none of us remember all of it, how we create the narrative of our life requires a degree of reasoning. "We remember events, and say, 'Well, that happened, so something else must have happened too.' We tell stories to ourselves that include real memories, plus things that we infer to have happened," adds Christoff. We are all detectives seeking to understand the story that is the ongoing mystery of who we are.

3

PASSION

In Cold Blood?

In *The Anatomy of Melancholy*, his astonishing Renaissance rant on the emotions, Robert Burton writes, "Perturbations and passions . . . though they dwell between the confines of sense and reason, yet they rather follow sense than reason, because they are drowned in corporeal organs of sense." As a cognitive scientist born centuries before his time, Burton was obsessed with the intertwined influences of passions, thought, and body. Now that the prefrontal cortex is beginning to reveal the nature of its powers of cognition, we can begin to understand the neural substrates of emotion and reason's complex dance of the mind.[1]

Living on Standby

In 2000, Steven Anderson and colleagues at Antonio Damasio's brain shop at the University of Iowa, compared personality transformations in prefrontal trauma victims. Did injuries to the dorsolateral and ventromedial PFC alter behavior in different ways? They also wanted to develop a better system to assess before-and-after personality metamorphoses. Psychometric instruments—such as the Minnesota Multiphasic Personality Inventory, or the Eysenck Personality Questionnaire—had failed to capture personality mutations in VM patients. Accuracy, furthermore, was at the mercy of patients' self-reports, and insight is a notable casualty of VM damage. "It just wasn't adding up," Anderson explains. "Descriptions from patients just did not match what we were seeing."[2]

Thus they designed their own instrument, the Iowa Rating Scales of Personality Change. Based on information supplied by family members and friends—those who knew the person "before and after"—it provided a superior index of personality change. This new study, furthermore, underscored the cardinal feature wrought by ventromedial damage: a subtle litany of absences. Ventromedial patients evince blunting of emotional experience and expression, weakening of initiative, and loss of persistence and decisiveness. They do not indulge in rash or impulsive acts so much as exhibit lack of judgment in matters that unfold over days and months.

Although none of the Iowa patients' behaviors approached wild, out-of-control dysfunction, they did react in emotionally inappropriate ways, at times displaying a fecklessness in the face of circumstances most people would find unsettling. When they did react emotionally, it often took the form of frustration, and they tended to explode over trivial setbacks. Unlike patients with dorsolateral PFC lesions, the VM group could reason just fine. Patients with dorsolateral damage, on the other hand, showed reasoning impairment but less of the VM patients' emotional voids and mental lethargy. So Anderson and his team became more convinced than ever that the health of the ventromedial region (which in their view includes part of the orbitofrontal PFC), with its rich interconnections throughout the brain, is a key mediator of emotional integrity and social behavior.

To picture what the ventromedial area looks like, imagine standing underneath the PFC and, looking up at the "ground floor," noticing its central seam running front to back (see figure 1 on page xi). The VM sectors hug this middle seam, while the orbitofrontal occupies the outlying cortex. Damage to the ventromedial PFC in even one hemisphere can throw a person's life into a shambles. Such a tailspin overtook a forty-year-old Danish man. Early in 1991, Copenhagen resident "LP," an MS in social science and communications, was enjoying a rewarding career teaching communications and information technology to young adults. He lived in a villa with friends and played guitar in a rock band. Although he was divorced, the split with his wife was amicable, and he shared custody of his two children and spent much pleasurable time with them. Then LP suffered a ruptured aneurysm of the anterior communicating artery—a major blood vessel servicing most of the middle portions of the frontal lobes, severely damaging his left orbitofrontal and ventromedial PFC, plus the anterior cingulate.

Although conventional psychological tests found LP to be normal in his reasoning capacity, as is so often the case after OFC/VM injuries, his life was far from normal. After convalescing for several months in Spain, LP returned to his job. But he "forgot" to prepare for classes, could no longer maintain his pupils' attention, and eventually lost control of the teaching process overall. He also failed to do his share in the villa, crammed his space with junk and garbage, and was eventually kicked out. Soon he went on the Danish equivalent of disability. Four years after the injury, he was retested for insurance/disability reasons and sent to Birgit Bork Mathiesen, who was then working on her PhD on personality change after brain injury. Struck by the contrast between LP's intellectual capacities and his real-life situation, Mathiesen offered him psychoanalytic therapy twice a week aimed at finding some coping patterns within his personality that could help him reconstruct his behavior.

A decade later, Mathiesen, at the University of Copenhagen, is still grappling with the meaning of LP's problems. With colleagues, she has written several reports on his progress or lack of it. When LP arrived at her office for therapy sessions, Mathiesen recalled recently, he seemed eloquent, smart, and capable of self-reflection. "When you were there, he was perfectly fine, intelligent, nice to be with. Yet he lived this terrible life. He could never get anything done. Neighbors complained because he didn't take the garbage out and the place smelled. He lived like a chronic schizophrenic."[3]

As LP was the first to admit, nothing affected him emotionally the way it used to, except music, which, he noted, he could still "get high" on. But his joy of playing alone or rehearsing with his band was gone. He played less often. Ignoring phone messages, letters, and appointments, he even forgot to pick up his children for their summer vacation. His pervasive mood during these times, reported

LP, was as if his emotions had retreated to someplace far away, as if his feelings were "on standby." When in this "arrested" mode, it was as if his thoughts stopped. Living on standby, Mathiesen told me, is an apt description of the status of people with ventral PFC lesions, especially those involving the ACC. "Often relatives complain about their strange apathy. Over and again, the person states his intention to do something; really, really means to get right on it, but never does," she explains.

Tests that link emotions to bodily responses—like sweating, muscle tension, and increased heart rate—showed LP had muted visceral reactions to emotional situations. This lack of "somatic markers" of emotion, a hallmark of the ventromedial-injured person, correlates not only with emotional processing but decision-making, as Damasio has eloquently articulated. "That's not to say LP was without emotions or empathy when you are with him," Mathiesen adds, still preoccupied with the paradox of her patient. "But his inertia just took over. He was aware of what he was supposed to do, ashamed of not doing it, but just not motivated to do it." Emotional limbo infected his ambition, drive, and ability to push on.

LP began every therapy session by announcing he had nothing to report. When Mathiesen reminded him that "anything that comes to mind will do," he presented "talks, like news shorts on TV: 'I went to the fitness club this morning and played table tennis, which was fine. Then I went to the movies; I took my bike. It is good that I got this bike.' And so on." This robotic recounting of events with a litany of factual detail instead of emotionally textured, empathy-arousing conversation, is something Mathiesen calls "pensee operator." "The speaker has difficulty distinguishing between different feelings and sensations and in verbalizing them."

Mathiesen once asked LP if he would talk about his life to a group of university psychologists. He readily agreed. "I interviewed him in front of this audience. He did fine and was quite intellectual about it. But afterward he said, 'Oh, that was hard!' And I said, 'I can imagine, being exposed to so many people.' And he said, 'No, no, not that. Getting all the pieces of my story to stick together—that was hard!' It was as if his autobiographical memory was split up in bits, fragmented. He couldn't summon up his past sequentially." The emotional glue that helps build the personal narrative wasn't there. Mathiesen suspects that LP no longer had an integrated picture of himself—or of others.

One of his close friends said what he missed most about the "new" LP was his "visionary attitude." LP seldom saw himself in the future, and no longer speculated on the meaning of life with his previous positive energy. Weakened, too, was his capacity to endure ambiguity or resolve conflicts between his dreams and the actuality of life. LP often chose superficial, quick payback over distant rewards. Like many orbitofrontal-VM patients, he was unable to focus and maintain emotion toward achieving long-term goals. And when he realized his mistakes, Mathiesen recalls, he would feel renewed shame.

LP's humor was weighted with sarcasm now. And while he realized his ill-timed and badly chosen jokes often cast a pall on the social ambience around him, he was incapable of suppressing the caustic commentary. Yet his friends thought LP was too hard on himself, and as they told Mathiesen, he seemed to feel more isolated from others than he in fact was.

"But perhaps his friends didn't realize how painful it was to have lost the richness of his emotional and social life," she suspects. "And he did, after all, lose a lot of friends. Maybe their reaction was, 'Pull yourself together, man. You don't have that much of a problem.'" But his social out-of-tuneness was inescapable. That LP was out of sync with others, Mathiesen theorized, might be the result of his injury disrupting the powerful broadband circuitry linking the PFC to its subcortical emotional information storehouses. "There are so many nonverbal cues we read without thinking of it. Maybe his lack of access to emotional information prevented him from making these split-second social adjustments."

Mathiesen treated LP over a two-year period. "In the beginning he was enthusiastic, but when the initial glow faded, he couldn't stick with it." He would disappear for weeks at a time, then return. He hooked up with a rehab team, but they too finally gave up on him because he blew off so many appointments. Could conventional therapy work for LP? "Certainly not the kind I thought he would profit from," she admits painfully. "But then I didn't realize how in-depth this lack-of-initiative problem was. I think he would need a person to come to his home a couple of times a week to get things done. He's not without conscience, so with someone to keep him involved he might regain contact with parts of himself that were there before. Maybe we could pull him in the right direction. But now, I think, he is lost. He cannot do it himself."

   LP's story reflects the difficulties in treating people with damage
to the emotional parts of the prefrontal cortex. Indeed, this ground
floor of the PFC is one of the most commonly traumatized areas in
head injuries, since it lies nearest to the nose and bony sinus areas.
No major cognitive deficits explain LP's problems: but this chasm
between emotion and the mental representation of content in his
mind affects everything he thinks and does. The Dane's dilemma
highlights the role of emotion in the mind's work space. The
VM/OFC must somehow make possible an emotional mental model
or script—a correlate of the dorsolateral PFC's cognitive working-
memory representation. One's own feelings, other people's emotions, social context—all provide an affective trellis on which we
maintain and support our personal narrative. Without an "affective
working memory" we cannot care enough about our remembered
"obscure objects of desire" to drive our executive processors to
achieve them in time spans short and long into the future.

   Conventionally, we regard the rational-emotional matrix of our
thinking with, well, mixed emotions. Emotion is often cast as the foe
of reason, yet since Darwin, some have proposed that emotions—
such as fear, greed, lust, and joy—facilitate the survival of species.
But what benefits do the passions confer on a species that has mani-
fold ways of deliberating and communicating? After blowing your
gasket for the umpteenth time, you might wonder if cognition
would be better served without emotions, if we had evolved, Mr.
Spocklike, to have shrugged them off as we did our vestigial tails.
And yet what important decisions in life does anyone make in an
emotional vacuum? What memories are devoid of emotional col-
oration? Is anything difficult accomplished without desire? Charac-
teristically dispassionate himself in his appraisal of the emotions'
potent symbiotic relationship with reason, Aristotle in the Rhetoric
offers that "the emotions are all those feelings that so change men
as to affect their judgments."[4]

   As brain researchers in the second half of the twentieth century
grew increasingly excited about the "hard science" of executive
functions, emotion was left to the psychologists. By 2000, however,
with a better understanding of brain chemistry, the advent of imaging systems, and more complex neural-net programs, affect began
to get a foot in the door of classical neurosciences. "A few years ago,"
reflects Todd Braver, "I might have said we can make much

progress without considering the emotions. But more and more I am growing to the view that when we disregard emotions and treat them as some extra variable we can dub in later, we run into brick walls. Emotions integrate different aspects of information processing all the time."

Because science places such a premium on objectivity, the notion that "pure reason" is actually deeply infected with emotionality might be hard for some researchers to swallow. Yes, *Homo rationalis* might say, there would be some points at which reason and emotion interface, but they're segregated systems, each doing its own thing. Wrong. "Sure, there is a lot of play and independence in the nervous system," says Braver, "and these systems are probably not fully integrated throughout the brain. But there is no dividing line where this side 'knows' what the cognitive input is, and that 'knows' about emotional input. They're wrapped together in many places." So in labs around the world, interest in the neural bases of emotion grew exponentially. When the SUNY neuroscientist Turhan Canli ran a PubMed literature search using the terms "emotion" and "brain," he found that the number of citations doubled between 1995 and 2000 alone.[5]

The Superhighway of Passion

Bushwhacking in a trackless forest in the dead of night, the blood races and sweat trickles down the spine. Our intrepid hiker feels emotions coursing through his body from head to toe. The next day, back in "civilization," the boss calls him into her office to announce that he is being promoted and rewarded a hefty bonus this year. His knees go weak. Perturbations and passions; suppressed hysteria and wild elation, indeed. The emotional neurosphere is, if not drowned in the corporeal organs of sense, tightly embraced with them.

Until recently, however, the brain-body circuits of emotional processing were known only in piecemeal fashion. It's taken over decade, but the Boston University neuroanatomist Helen Barbas has fairly definitively mapped these large pathways of the brain-body axis as a neural interstate coursing in a powerful and direct line between the PFC and lower organs. Working with monkeys, she discovered that neurons originating in orbitofrontal and ventromedial PFC structures (including the ACC) transmit signals both about one's internal, bodily milieu and the external world on express routes running down the body.[6]

This neural turnpike that runs like I-95 down the East Coast has but one major stop between the frontal cortex and the brain stem. It is the hypothalamus, which like the Washington, D.C., beltway or a major Internet hub, has extensive nodes and connector roads that tie into basic functions, what the neurologist and author Alice Flaherty call the Four F's: "fear, food, fighting and . . . sex."[7] These interlinked beltways are also yoked to emotion's Times Square, the amygdala, which sends communication tentacles all over. Then, like I-95 continuing south to Florida, the superhighway continues downward, connecting to brain-stem and spinal-cord autonomic systems. These hookups enable fluctuations in emotional states to affect heart rate, oxygen use, blood pressure, muscular contraction, and glucose metabolism by directly and indirectly stimulating all bodily organs. The expressway provides the prefrontal executive an efficient mode of response to complex emotional situations. The PFC systems evaluate the significance of an emotional situation, and the subcortical structures, having gotten the signal from above, express the appropriate emotion through visceral changes in the rhythm of the heart, the adrenal glands, and the sweat glands.

There are subtle differences in the orbital and medial, outer and inner PFC connections. The orbitofrontal cortex shares with the amygdala "a panoramic view" of the entire sensorium through robust ties with the domains of touch/pressure, hearing, vision, smell, and so on. Thus your OFC has double access to your sensory world: directly from the sensory brain, plus an emotional track from the amygdala, enabling the OFC to extract and weigh the emotional tenor of events beginning at the periphery of your body. In contrast, the inner medial PFC shows comparatively sparse links with the external world—it is the more abstract, self-to-self arena. The medial PFC has dual tracks to and from the limbic systems that also interface with the hypothalamic and brain-stem autonomic centers.

Awareness of emotions requires the higher cortex. The amygdala, however, fires even when people are unconscious of having glimpsed pictures of fearful faces, being hot-wired for vigilance and fast reaction when danger lurks through lower-order pathways that bypass the cortex. Hearing the castanets of a rattlesnake requires no cortical help to raise your blood pressure. But for acute awareness of the emotional impact of this snake in your vicinity, it's probably necessary to mobilize pathways linking the amygdala and orbitofrontal PFC. The OFC and the medial PFC are tied to each other as well as to the more "intellectual" lateral PFC areas. In concert, these PFC structures can evaluate emotional events and key up the lowest levels of the nervous system.

The Asymmetry of Emotion

The home page of Yale's SCAN Lab features a Venn diagram—those three interlocking circles, like the Ballantine beer logo—in which the truly powerful stuff happens in the overlapping sweet spots of the circles. In SCAN Lab's Venn, one circle represents emotion; another, individual differences, including personality and intelligence; the third, cognition. The sweet spots, however, hold question marks. And it's what's happening at the nexus of emotion, individual differences, and cognition that SCAN Lab explores.

"The question marks," says Jeremy Gray, the director of the SCAN Lab, whose acronym stands for Social Cognitive and Affective Neuroscience, "are the areas that have fallen between the traditional lines of inquiry. People have always talked about individual differences in emotional expression as being strongly related to personality for a couple thousand years." But where do they all conjoin? Gray's investigations pinpoint for the first time the prefrontal cortex as the region where the spheres of emotion, intellect, and personality intersect.

But first there is the sprawling, unruly scope of "affective" brain science. "'Emotion' is a huge, huge term, as big as 'cognition,' and the concepts are just as slippery," admits Gray. "It's easy for people to talk past each other unless they agree on the meanings." By 2003, though, the field had set up a few ground rules. First, there are no isolated joy, rage, love, or hate "centers" in the brain. Emotion is the outcome of a set of components interacting within a distributed network of brain circuits. And emotions, as Barbas and others demonstrated, reside not just in the head, but involve the whole body. This body connection is a big component of "feelings," as Antonio Damasio describes in his popular accounts of the "somatic marker" theory. Nor are emotions necessarily conscious, "supraliminal" states. Much of our emotional life is subliminal, beneath the surface of awareness, "down in the engine room," as philosopher of the mind Daniel Dennett puts it.

Parallel to understanding how emoting fits together with reasoning, affective explorers also confront the opposite puzzle: how to disentangle the myriad, overlapping emotional processes into more elementary parts.[8] Many affective neuroscientists are like modern-day Richard Burtons equipped with clanking MRI tubes. Like the seventeenth-century polymath, they seek a precise anatomy of melancholy, anger, pleasure, disgust, desire, and so on. But whether someone is parsing the neural correlates of greed or exploring the integration of joy in abstract calculus, one notion serves as a springboard: hemispheric asymmetry.

At the cartoon level, hemispheric asymmetry theory posits the left PFC to be the happy lobe, while the right prefrontal lobe is the trauma center of sad and scary emotions. Until recently, EEG findings implied a kind of good-humor homunculus residing in the left PFC, and a troll-like malcontent glowering over in the right. A psychologist would compare the balance of resting activity in both of a person's PFC hemispheres to see which critter dominated and thus obtain a measure of the person's "disposition." As such, electrophysiological recordings of a baseline of healthy individuals added weight to an affective yin and yang theory of emotional PFC operations.

This scheme grew out of the older conceit of hemispheric balkanization: that the right hemisphere specializes in all emotional processing, while the left is the realm of cold, logical calculus. This idea is rooted in nineteenth-century observations that patients with right-hemisphere injuries seemed more apathetic and passive, while those with left-hemisphere damage were typically more emotionally volatile. This right-side-as-emotional-side theory floated around for about a century, and in part inspired the "drawing from the right brain" cliché that is now embedded in pop culture—that the right hemisphere is the seat of "creative" impulses, while the left is a deliberation machine, cranking out rational ideas.[9] The case for left-right emotional specialization arose in response to later, still fairly primitive, studies of patients with damage in one hemisphere or another. People with right frontal lobe lesions were prone to pathological laughing or bursts of euphoria, those with damaged left hemispheres to crying spells and depression.

Meanwhile, to trigger select emotions in lab subjects, psychologists amassed a catalog of powerful props and scenarios. These compilations, included in the International Affective Picture System, generally fall into three basic categories: pleasant (such as embracing/erotic couples, bucolic landscapes), neutral (house and garden scenes), and unpleasant (attacking animals, menacing humans, visages frozen in rictuses of surprise and horror). To this Emoting Family of Man have been added arrays of film and video clips ranging from saccharine narratives and comic films to war scenes, grotesque mutilations, amputations, and filthy toilet graphics. Ever-ingenious experts has also rolled out depictions of tempting foods and gambling games with cash incentives—techniques that reliably measure how you calibrate short- and long-term reward versus the threat of money losses. Studies of people being hot-buttoned into extreme mood states may seem laughable, yet they can be near-torturous to volunteers, who like actors in TV survival show episodes sweat it out for their white-coated audiences.

By the late twentieth century, Richard Davidson had begun to put his stamp on a growing mountain of affective data. A prodigious worker with an acute sense of public relations, the University of Wisconsin psychologist developed his own version of the hemispheric asymmetry model called the valence hypothesis. With his curly hair and long face, Davidson seemed to pop up everywhere. For a time, it was hard to miss media coverage of his touted exploits, notably experiments on the mood-altering power of meditation techniques. Davidson, who works primarily with the venerated EEG technology (seemingly as old as meditation itself), has hauled his equipment Herzog-like across mountainous Tibet to measure the brain waves of Buddhist monks while they are meditating. He also claims to have enhanced the immune response of biotech workers by having them practice meditation for six months.

The "valence" in the valence hypothesis refers to the attractiveness or repulsiveness of goals. There is an implied physical dynamic—emotion as directed motion, as vector. A happy goal is said to have a positive valence: this feeling draws you forward as if by magnetic attraction. A negative goal is unattractive, repellent, threatening, hence it has a negative valence. The hypothesis also proposes that the prefrontal hemispheres are "lateralized" to handle either negative or positive emotional valences.

Some imaging experiments support Davidson's notion. Japanese scientists scanned volunteers who were warned they were about to see emotion-soaked pictures. When a subject expected to view a gruesome picture, the right orbitofrontal and medial PFC lit up along with the amygdala and ACC. When the subject expected to see a pleasing image, the left dorsolateral and medial PFC lit up. This suggested that when you anticipate a specific emotional valence, prefrontal areas "prepare" you, via the "appropriate" hemisphere. In another study, seventy-two heavy smokers were asked to lay off cigarettes for twenty-four hours. The next day, the nicotine-famished volunteers were put in the scanner immediately after they were told they could smoke again in two minutes. The positive, left PFC blazed emphatically as the relieved subjects anticipated their cigarettes.[10]

More precise neuronal studies give further credence to the hemispheric asymmetry theory. In a rare example of single-cell recordings in a human, Ralph Adolphs monitored single neurons in the right PFC of an awake patient awaiting epilepsy surgery. Adolphs, now at Caltech, found that these right PFC neurons buzzed when the patient viewed grisly pictures of mutilations, war, and other disasters. Sunny or neutral pictures did not elicit the same response. The nattering neurons, furthermore, spiked within about 0.12 (one-eighth) second after the patient laid eyes on the aversive pictures, probably before he consciously saw them. The cells' rapid-fire behavior is consistent with the brain's need to respond swiftly to sudden, potentially dangerous events.

In another study, Damasio's team recorded from neurons in three surgical patients, this time in both PFC hemispheres, while the subjects watched movie clips depicting happiness, fear, anger, disgust, or sadness. Multiple electrode contacts permitted recording from nearly one hundred neurons per subject from several separate sites. Neurons in the right PFC fired for fear, disgust, and sadness; cells in the left PFC fired only for positive images.[11]

The emotional asymmetry literature is nonetheless replete with exceptions. Anger, for instance, seems associated with the "positive" left PFC rather than the negative right hemisphere, suggesting that an emotional vector in the PFC, or some kind of force field, is associated with the direction of your anger. And brain sites excited by fear differ from those elicited by disgust. Today, scientists utilizing imaging systems are compelled to move beyond the simplistic categorizations of emotion as right or left PFC-activating, or as either positive or negative.

Jeremy Gray, who once "was very taken" with the model, has worked to clarify its status. But in fMRI studies, it turns out, emotional asymmetry is not as starkly delineated as certain cognitive divisions of labor. A recent meta-review of all imaging experiments found weak fMRI support for the emotional lopsidedness so prominent in the EEG studies. "That there is this clear marker, this asymmetry," Gray grants, "is a very sexy hypothesis. And there is some tantalizing evidence out there. I still think it's a neat idea and would be as delighted as anybody if there was strong evidence for it. But I'm reluctantly coming to the conclusion it's just not bombproof, like cognitive asymmetry. It's not like the left prefrontal cortex is active only when the right is not. There's something pretty subtle, and you don't always find it."

Affective Style

That each person has an emotional personality, stable over time, is an assumption central to the hemispheric asymmetry model. While the model remains empirically ambiguous, it's clear that everybody has a unique affective profile, with unique balances of positive and negative emotional sensitivities. Indeed, a key to understanding the PFC involves unmasking the neural underpinnings of this "affective personality bias." A longtime staple of women's magazines, personality-profiling quizzes and emotional IQ tests now proliferate wildly online, from "Who's Your Inner Rock Star?," "The Slut Test," and "What Kind of Underwear Is Right for You?" to "What Breed of Dog Are You?" to the mundane "Should You Go to Graduate School?" to the classic "What Color Is Your Personality?"[12] Personality tests used by psychologists are only slightly more technical.

For some researchers, affective style determines temperament. In his quest for a neurobiology of personality, Richard Davidson posits emotional style as a "major ingredient for many fundamental dimensions of personality."[13] According to affective style theory, moreover, your disposition incorporates two "systems" of emotion that operate simultaneously. One system is akin to extroversion—

upbeat emotions, such as enthusiasm, joy, pride, and pleasure. This is the so-called Behavioral Approach System (BAS), and it purportedly facilitates a "moving towards" impulse, such as outgoingness and gladsome emotions that occur as one aims for a desired goal. The other dimension, the Behavioral Inhibition System (BIS), resembles introversion, and encompasses susceptibilities to fear, disgust, and increased vigilance that foster a "getting the hell away" desire to retreat. These counterposed systems are putatively implemented in partially separable pathways whose key components are the PFC and limbic areas, although other brain areas are also implicated.[14]

The BAS/BIS theory dates to studies by Neal Miller sixty years ago. A pioneer in biofeedback studies at Yale, Miller actually tethered rats in tiny harnesses and measured how strongly they would pull in one direction to approach or another to withdraw from a stimulus. Approach/withdrawal vectors, then, appear more empirically grounded than vaguer measures like "positive" or "negative," and it's easier to make behavioral interpretations about them. "If you can say, 'Okay, the person, or rat, is taking a step forward or backward,' there is a clarity to it," says Gray, "and an emphasis on the goal-directed nature of these emotions." BAS/BIS dimensions are good descriptors, he says, because they define action-oriented emotions, "urgent neural sparks that can rapidly turn into fires." The emotions are not simply sitting there, they are urging you to do something. Goal-directed emotions might interact, interface, and coordinate with cognitive systems in your behavior about goals.

Several interpretations of the BAS/BIS model propose that your affective style is a consistent, if dynamic, mix of approach/withdrawal systems; your unique ratio of BAS to BIS determines your personality. Although each dimension operates somewhat independently, being partial to rewarding experiences does not rule out susceptibility to withdrawal, negative emotions. You can score high, or low, or somewhere in the middle on both BAS and BIS. Personality, then, is a kind of existential "place" where your coordinates lie along the possible dimensions. Designate BAS and BIS as X and Y axes, and Z as one's mood at a point in time, and the system becomes a kind of Cartesian axis of personality. (This construct is not unlike myriad biochemical ratios, such as "good" and "bad" cholesterol levels.)

Individual BAS/BIS scores in EEG studies predict for differences in prefrontal asymmetries. In one test, high-BAS scorers had greater left PFC activity at rest—when their brains were basically idling—than did high-BIS scorers. Conversely, high-BIS participants had greater right PFC activity under the same conditions. Another experiment comparing extremely extroverted (BAS) and introverted (BIS) women showed the hypothesized hemispheric "lopsidedness." But don't plan on getting an EEG that will reveal you to have an extreme BAS or BIS personality profile. Most people, whose EEG profiles are somewhere in the middle, will not show this clear-cut asymmetry.

Some investigators view affective bias as a constant across a person's life span, starting from infancy. They've used a tangle of EEG electrodes embedded in Buck Rogers–like helmets to measure electrical patterns from the scalps of ten-month-old babies. Davidson claims that infants with greater right PFC activity are more likely to cry in response to a brief period of separation from their mother than are laughing babies who showed more left PFC activity. Toddlers and young children with greater relative right PFC activation show more shyness and wariness than left-PFC-dominant children, who are more gregarious. In a study of ninety adults, Davidson segregated people whose left PFCs fairly screamed with activity from those whose right PFCs were similarly energized. Then he asked these two divergent groups to choose from a list of positive and negative adjectives the words that described their pervading emotional disposition. Those choosing adjectives such as alert, strong, excited, proud were those whose left PFCs were jazzed. People selecting adjectives such as distressed, nervous, and scared scored high on right PFC activity.

How is the prefrontal cortex the axis of "affective style"? If the data can be trusted, the very code of one's personality emerges from the PFC in concert with affect-dedicated lower-brain partners. This asymmetrical balancing act in the prefrontal cortex may regulate the intensity of your affect—turning emotions up or down—as well as compute the emotional value of events, actions, and mental representations. The PFC may well be the architect of "emotional working memory," that process (seemingly lost to the Dane, LP) that holds in mind the representation of an emotion in the absence of immediate rewards and punishments. Theoretically, one's "affective setpoint" will endow memories with enduring positive or negative force.

In its role as planner and anticipator, the PFC will emotionally tint the "remembered future" as well. Those of us displaying dominant-tending left PFC activity may envision a more pleasant tomorrow than those with a more right-dominant PFC who may dread what's around the corner. There is a form of happiness Davidson terms "pregoal attainment positive affect" that arises when you anticipate a happy outcome down the road. In a prisoner-of-war situation, one captive might be able to withstand more pain than another because he can sustain a mental representation of a positively charged future he will fight to live for.

Conversely, if someone is prone to fearfulness, his or her future plans might be suffused with warning signs and caution, a "pregoal attainment negative affect." This, too, has adaptive value. A little paranoia can come in handy. We should heed our negative prophets; the human race needs representatives "with their hair on fire." Purebred left-PFC-dominant optimists might not think it necessary to design sufficient contingency plans for errors, breakdowns, and more disastrous things in air- and spacecraft, buildings, and financial budgets. And yet excess anxiety turns off reasoning processes, and without pie-eyed optimists there is little future advancement.

The Grouch Factor

Personality may also spring from activity-level variation in specific substations of the prefrontal cortex—indeed, these variations may define personality. Take the ventromedial PFC (BA 25). Lying smack up against the inner medial wall just above the PFC floor, the postage-stamp-size VM is so deep within the PFC that EEGs can't record from it at all. With fMRI being a less precise technology than PET, Vanderbilt's David Zald and colleagues went after the VM with the older instrument. And in a study of eighty-nine healthy volunteers, they saw differences in activity there correlated neatly with how much negative affect each subject had—a package including pessimism, anxiousness, worry, and downright orneriness. The more "worked up" the VM when the subject was in a resting state, the more likely he was to report frequent bad moods. In other words, high-arousal VMers were a grouchy, irritable lot.[15]

But what comes first? Where is the alpha point in this emotive circuit that determines personality trait? The VM, Zald thinks, is not secondary to what's going on elsewhere, but is "critical to modulating the severity of the negative affect." It functions as the dial that ratchets up or down the intensity of uneasy passions, and this volume control is what we interpret as a temperament. Thus people with higher-than-average ventromedial activity might react with more nervous intensity to events than those with lower VM activity who experience less angst in even sharply jarring experiences. Someone's lower-than-normal ventromedial set point might be the basis of his or her craving for scarier experiences, such as scaling the rock faces of El Capitan or bungee jumping. Conversely, if one has a higher VM set-point, he or she might prefer to go to the mall in a Hummer.

The ventromedial PFC anxiety thermostat may also help modulate the time course of anxious events, how long it takes to return to your affective cruising speed after a jolt of nerves. This VM activity set point may well be something a person is born with. "My gut hypothesis," says Zald, "is this is a biological variable that may start very early as a temperament issue. We know some babies are 'colicky,' some more reactive than others, and there's good evidence that there is continuity of these traits." The degree of VM activity may then be a heritable trait—one predictive of a person's general level of negative affect. "It's a simple, easy, straightforward hypothesis," Zald adds, "but there's no ethical way to test it. We can't take one- or two-year-old kids and scan them with PET."

As the VM is intimately linked with the body's alert systems, its overarousal can have negative health consequences, such as high blood pressure and stress-related disorders. People with negative emotional biases tend to have more divorces, alcohol abuse, and hair-trigger reactions that don't endear them to polite society. Understanding the relationship between ventromedial PFC function and mood will lead to more effective treatment—both pharmaceutical and therapeutic—for depression, anxiety, and other "psychosomatic" disorders.

There are, again, survival advantages to being tightly wired and angst-ridden. One high-strung New Yorker, for example, is so worried she will never find a taxi that she has turned herself into a competitive sprinter. Not everybody will notice or respond to something subtly dangerous in the environment. The curmudgeon may be less likely than the blithe spirit to ignore vaguer dangers. "If there is an edible delicacy or a beautiful person out there, yeah, it's nice to attend to it, but if during a crisis we don't, we're not going to die, we'll live to see another day," adds Zald, noting that some self-help books actually promote getting in touch with our inner fear. "If you meet someone and for some reason feel anxious near them, for instance, why risk being around them?"[16]

Zald adds support to a feature of the BAS/BIS model: negative is not precisely the opposite of positive affect. Having one's ventro-medial PFC chemically suppressed, say, will not make you happier. People such as patient LP with diminished VM capacity often strug-gle to weigh risks and are immune to emotions and physical sensa-tions of worry. Such a person, says Zald, might consider betting $10,000 on the roll of dice and not have the slightest twinge of nerves. A friend could warn him that his chances for failure were excellent and he'll have no visceral or emotional reaction to it, even though he swears he understands the odds are stacked against him. In a study now canonical in the affective literature, the Damasios and Antoine Bechara measured the body reactions of patients with ventromedial prefrontal lesions and control subjects while they played a computer gambling game, the Iowa Gambling Task. VM patients kept playing losing decks until they went broke.[17]

Zald's further studies indicate that differences in this anxiety index may affect how people think and act in emotionally intense situations. His group found that a person's ability to control his or her attention was directly linked to his or her personal anxiety lev-els. In a test situation, more laid-back people are better able to stay focused on targets while also being subjected to emotion-fraught images than those with high harm avoidance scores. The high-anxiety folks may have more trouble disengaging from, or sup-pressing, emotional images than their more insouciant counter-parts, causing their attention to stay locked on an emotional image even when it is of no value to do so.

Special Forces

Resilience in response to stress may also be a partly innate person-ality trait. Although most studies scrutinize the aftermath of a stress-ful experience, a Yale psychiatrist explored differences in how individuals respond before, during, and after stressful action. Charles "Andy" Morgan III, working with investigators at the Department of Defense, discovered that levels of a hormone and transmitter, neuropeptide Y (NPY), may be largely responsible for how well an elite warrior functions under the pressure-cooker conditions of warfare, and how quickly he or she recovers from ordeals of combat.[18]

"People who release high levels of neuropeptide Y under stress stay mentally focused longer," says Morgan. Special Forces candidates, he found, mobilized significantly more NPY than did other soldiers, including Rangers and Marines, who were undergoing the same training. The commandos sustained better navigational skills, stayed calmer during potentially life-threatening events, and committed fewer errors in the fog of simulated war. After enormous physical and psychic wear and tear, they experienced less burnout, bouncing back much faster than those who had not released the same levels of the hormone. Among other functions, Morgan thinks, NPY works in the PFC to help a person maintain mental clarity, think fast on one's feet, and suppress many physiological distractors brought on by high stress.

With far-ranging receptor sites throughout the bodies of most animals, NPY helps the sympathetic nervous system perform balancing acts of heart rate, temperature, appetite, and fight-or-flight impulses. NPY puts the brakes on the stress hormone adrenaline and the transmitter norepinephrine, so that the sympathetic nervous system doesn't overshoot its optimal alerting state. Low baseline levels of NPY may prove to be a key indictor of vulnerablity to posttraumatic stress disorder.

Although for decades the military has profiled its commando recruits to predict who has the emotional right stuff, nonetheless nearly four out of five candidates drop out. The armed forces wanted a way to prejudge who can handle counterterrorist ops and other extreme challenges and return home without mental impairment. Morgan studied dozens of GIs chosen for Special Forces during a training period at the JFK Special Warfare Training Center and School in Fort Bragg, North Carolina, while they confronted a regimen involving night hikes, immersion in frigid water, disorientation, simulated capture, interrogation, and torture based on real prisoner-of-war experiences. "The candidates were compelled to confront a regimen of stressors," says Morgan, "which, if applied in real-world operations against detainees or POWs, would exceed what is permissible under Geneva convention standards. Here, of course, it is training. And the soldiers volunteer to undergo such stress in the hope that they will be selected to be a member of these elite troops."

In evaluating the candidates physically and psychologically before, during, and after the training, Morgan's group found that variation in individual performance correlated strongly with NPY levels. At the outset, every soldier showed similar levels of stress hormones and NPY. During the height of the ordeal, everyone initially had powerful reactions to the stress. Their levels of hallmark stress hormones—epinephrine and cortisol—spiked to twenty times normal; that is, to levels seen in patients undergoing heart surgery! And everyone's NPY levels rose as well. Then the defining differences emerged: some men's NPY levels soared; they turned out to be the soldiers who performed best and recovered fastest afterward.

The day following the trial's end, most soldiers' stress hormones remained elevated, while their NPY levels were below baseline. They were depleted, still feeling the lingering trauma. Successful Special Forces candidates, however, were rested and ready, and their protective NPY levels had returned to baseline within twenty-four hours. A clear relationship also existed between the soldiers' NPY levels and their cognitive state: the more spaced out and disoriented they were from "dissociative symptoms," the lower the soldiers' NPY levels. And the more spaced out, the worse their training performance.

Successful commandos are not soldiers who feel stress less than others, Morgan explained. In some ways they may be even more sensitive than most people. But they produce higher levels of NPY that act as anti-anxiety agents and facilitate executive functions, such as rapidly crunching lots of information and synthesizing it into successfully decisive acts under duress. One wonders if this is the basis of the "great performer's syndrome"—whether soldier, concert pianist, or big-game athlete? The elite performer feels the jitters at least as much as the lesser performer, but rather than being paralyzed by them, is stimulated to greater feats.

Which came first: elevated NPY, or something that helped successful Green Berets be more confident, perform better, and thus produce elevated NPY? That one can train to become more stress-resilient is, of course, the postulated purpose of boot camp and advanced exercises. Certainly the memory of prevailing over adversity will help mitigate a person's anxiety in facing future hardships. In a broader sense, well-nurtured childhoods may lay the foundations for resilience. But the question remains: are some people born resilient? Perhaps. Correlating hormone levels with the prefrontal effects, Special Forces commandos with high levels of NPY tend to score above average on g intelligence tests, as well as lower in negative personality profiles.

Morgan thinks a beginner's future performance is predictable, based on NPY. His team replicated their findings in both men and women at the Navy Combat Dive School in Key West, Florida. Some Special Forces units now go to NIH for further tests of this resilience quotient under the direction of NIMH's Dennis Charney, who is also studying other soldiers and some prisoners in the Vietnam War. The hope, he says, is to find out how they handle stress differently. "If we could train people to mobilize their own neuropeptide Y, that would be a primary approach," says Morgan's colleague Matt Friedman, the director of the U.S. National Center for Post-Traumatic Stress Disorder.

Sex and Sex Differences

If individuals are unique in their emotional PFCs, are there verifiable sex differences in emotional brain processing as well? Conventional wisdom has it that the affective lives of men and women differ in everything—perception, experience, and expression of emotions. Certainly aggression is greater in males, and the intensity of just about all other emotions seems greater in women. The first wave of brain imaging studies, however, is showing fewer differences than one might expect, especially in the PFC.

At the Neuroscience Research Center of the Université de Montréal, Mario Beauregard wanted to see just how affective wiring differs in men and women. So he compared how each sex reacted to three different emotions. He scanned twenty men and twenty women while they watched three kinds of video clips: an episode of the British comedy series *Mr. Bean* for amusement, scenes of bodily mutilation to elicit aversive feelings, and pornographic film clips to excite sexual arousal. For a neutral control he used a video demonstrating carpentry skills—guaranteed to cool the perfervid forebrain. Participants rated the intensity of their emotional reactions on a scale from zero to eight. And for the most part, the neural centers turned on by each kind of movie were identical in men and women: the amygdala, the hypothalamus and thalamus, and the medial PFC.[19]

But the women reported their negative reaction intensity higher than did men for the mutilation film. The men rated their sexual excitement intensity higher than did women for the sex clip. And neurally, Beauregard also saw a "perfect correlation": the more intense the participants judged their levels of emotional reaction— either in dislike or sexual arousal— the greater the extent and the number of brain zones lit up. *Mr. Bean* amused men and women to the same degree, and brain activation patterns likewise showed little difference. Neurally, the women's brains fired more intensely during the mutilations.

But the question remains: could this be a result not of intrinsic neural wiring but of cultural conditioning, whereby men are inured to violent images? And likewise the men's reaction to the porn imagery: cultural conditioning? Neither gender difference may be inherently biologically based, Beauregard replies. A cultural convention could be processed as a neurobiological event. The erotic material of a pornographic film, designed by men for men, could hardly be expected to turn on the women, I persist. "We did take this fact into consideration," Beauregard admits, hastening to add that he plans a study that exposes women to other types of erotica. By looking at twins and people of different cultures, he also hopes to tease out genetic and cultural components of these differences.

While countless previous studies support the theory that men typically experience greater sexual arousal from visual erotic stimuli than women, little was known about the neurobiological processes underlying this gender difference. It's no surprise that when both sexes viewed erotic clips their brains' visual areas were busier than when they watched the carpentry films, as were many subcortical circuits associated with emotion. In two crucial subcortical areas men and women differed: men showed greater reactions in the thalamus and hypothalamus, some of whose sexually differing sectors are known to play pivotal roles in physical arousal, sexual orientation, and behavior. The magnitude of this primitive-brain activity paralleled the men's personal rating of how aroused they were.

Imaging studies of sexual emotions are surprisingly uncommon, and the first PET scans of men's brains at the point of ejaculation have only recently been announced. Dutch neuroanatomists found that one of the most active regions of the male brain at the point of orgasm is the ventral tegmental area (VTA), a midbrain cell cluster whose principal neurotransmitter, dopamine, plays such a central role in reward and euphoria. Studies of drug users show that the VTA is a major player during a heroin rush. Conversely, other neural orchestrators of emotions are turned off during orgasm, notably the vigilant amygdala. The researcher Gert Holstege thinks the amygdala's suppression is related to what most people know: that when we are sexually high, the outer world and its cares fade away. Passionate lovers show PFC activity levels approaching zero. Sexual ecstasy is not a cognitive state, and strategic planning, to put it mildly, is not the mental focus of the orgasmic animal. In a similar but preliminary study, Holstege's group scanned women reaching orgasm and found their brains were afire in patterns similar to the men's.

If subcortical function differs to some degree between the sexes, so, too, does the anatomical architecture. Ruben and Raquel Gur at the University of Pennsylvania Medical Center used MRI to measure differences in over fifty male and fifty female brain volumes in emotional limbic and prefrontal areas. Men and women, they found, had identical volumes of amygdala and hippocampus as well as dorsolateral PFC. But the women had larger orbitofrontal cortices, resulting in a substantial difference in the ratio of OFC to amygdala gray matter volume. This larger volume of orbitofrontal cortex "may relate to behavioral evidence for the sex differences in emotional processing," the Gurs note. Since the OFC—that emotion/reason interface agent—may be more adept in shaping women's analysis of emotional events and in emotional working-memory operations, women show better memory recollection for emotion-laden autobiographical events. The Gurs' work supports a theory that men and women arrive at their temperamental differences by using somewhat different neural processes. One might infer, then, that men are more driven by their subcortical limbic system than women, while women may have a more cognitively weighted emotional balance.[20]

The Catalog of Passions

Fear and trembling: in the catalog of passions, perhaps the primal emotion is fear. To understand prefrontal processing of this primal feeling, experts often examine how fear operates in people who can't control it. Richard Davidson, for example, devised a study of social phobics—specifically, people terrified of speaking before an audience. In terms of ingenuity, this experiment ranks just below the Yale psychologist Stanley Milgram's legendary work in the early 1960s.[21]

In "While a Phobic Waits," Davidson concocted a scenario to ascertain how right-hemisphere PFC activity paralleled bodily signs of stress when phobics were provoked to high anxiety.[22] Because people with inordinate dread of public speaking are especially prone to the physical signs of stress, Davidson recruited these phobics, who went off medications prior to testing. He assembled a requisite control group as well. To start, the phobics were told they would be delivering a short speech to grad students and research scientists who had "an interest and expertise in interpersonal behavior." Half the audience would sit before them; the other half would observe them from behind a one-way mirror. This expert audience, the phobics were informed, would not only rate their presentations but also their "general personalities." To jack up the tension still higher, the speech topic would remain a mystery until one minute before the stage call. During the waiting period, the phobics listened to a recorded message that every half-minute informed them about the remaining seconds to showtime. Measurements of bodily stress signs and EEGs were taken.

When informed of their speech topic—a current political issue— the phobics were then told, "Wait, change of plans!" They'd have not one but two minutes to prepare. Again the recorded-voice countdown began. Finally they were led to a small auditorium. At this point, the speakers were given yet another time delay. Ever attentive to detail, Davidson had his faux experts dress in white lab coats, and had them enter marks on rating sheets while the subjects spoke. When at last they finished their speeches, the subjects returned to the lab, where EEGs and autonomic measurements were taken again. Perhaps the phobics were biting their nails more than most, but it's doubtful many people would emerge from this challenge without some autonomic spiking and a bit of withdrawal-emotion processing in the right PFC.

Not surprisingly, the phobics recorded higher physiological markers of anxiety at each stage of the experiment than did the controls; although the controls' heart rates shot skyward during the anticipatory phases, it was not as much as the phobics'. The most pronounced divergence in body signs happened when the subjects anticipated delivering their speeches. EEGs showed that the phobics' right PFCs buzzed much more than the controls' as they awaited going on stage. Obviously, thinking about the speech precipitated more withdrawal emotions for the phobics (who probably wished they were anywhere else). The controls had more left prefrontal activity than right, perhaps reflecting positive approach emotions inherent in their planning, perhaps an anticipatory eagerness, rather than the negative dread felt by the phobics.

Fear, dread, and their incendiary sibling, terror, are yoked to the anticipation of pain, distress, and grief. Fear's aversive anticipatory power enables this dark emotion to twist and warp supposedly well-reasoned perceptions and decisions. Certainly fear is a powerful instrument in some politicians' toolboxes, marshaled to drown out rational, nuanced analysis with knee-jerk responses to alarms and threats. Thoughts about money, too, inspire a form of dread called loss aversion—an irrational mental calculation that elicits the negative feelings from the anticipation of a $100 loss, say, more acutely than the positive feelings from an anticipated $100 gain. Thus money calculations can evoke powerful withdrawal emotions, even as they arouse fear's approach partner, avarice. Indeed, a Wall Street cliché holds that the stock market is fueled by dueling fear and greed. The former hedge fund operator, financial blogger, and star of CNBC's *Mad Money* Jim Cramer has recounted the viselike hold loss aversion had on him. During the last year of his fund, he realized losses of $425 million and gains of $575 million, meaning he netted $150 million: a great year by most standards. Yet he was obsessed by the losses. Finally, after he had nearly destroyed his marriage and relationships, his family, friends, and business partners staged an intervention. He went cold turkey on stock trading the next day.[23]

But fear, like anxiety, can have protective value. Ventromedial lesions seem not so much to impair one's response to immediate threats or rewards, as to diminish or destroy the anticipation before, and immediate memory after, the emotional event. The ventromedial PFC may use working-memory ensembles to build mental models of enduringly positive states, even in the absence of enticements—and the same for negative, fearful states. The neighboring orbitofrontal PFC, on the other hand, may act in rapid "real time" learning of good/bad, fearful/desirous associations, and the "unlearning" of those associations.

Of course, it's difficult for the brain explorer to know precisely what mood a subject is truly bringing to the experimental table. Adrian Owen has tried to address this ambiguity in designing his gambling experiments, yet still isn't sure of the clarity of his findings. In spite of knowing that putting a chip on black has better odds of winning $100, Owen says, "we might for whatever arbitrary reasons decide we are going with red anyway. So that decision becomes a higher, lateral PFC function. In a sense, it gives you the freedom to do almost whatever you wish, rather than what you ought to do if you want to win." It returns us to the question of the role of other "irrational" emotions in the calculus of decision-making and "free will."

Owen considered his subjects' strategic deliberations about pounds sterling and included comments from people performing the gambling task. "They dream up reasons why they're winning," he says. "'For sixty pounds, I go left as long as there's a forty percent chance of winning.'" So does every bettor at every casino and racetrack in the world have flaming hot orbitofrontal cortices? "Ha-ha. I'm sure," Owen responds. "Because if you look at behavior, for a large portion of the time we optimize our performance to gain reward. Yet we don't always do this in a conscious strategic way. I may choose to take on jobs that will earn me more money. But for the most part, our behavior is fairly unconscious, not all thought out. The fact that some of us do very well, earn lots of money—all these rewarding activities are not necessarily conscious."

Anger

The *Iliad* was composed in the keys of anger. Human culture is infused with it. Anger is a "perturbation that carries the spirits outwards," notes Richard Burton. Volcanic wrath, fury, ire: anger is an approach affect whose neural correlates are unique among emotions. As adherents to the valence hypothesis have noted, anger manifests itself differently than negative/withdrawal emotions, since it seems to be a more "left" than "right" brain emotion.

Psychologist Eddie Harmon-Jones, now at Texas A&M University, wondered if this left PFC activity would be greater when angry people believed they were empowered to do something to extinguish the source of anger, a tactic called "coping potential." He compared PFC activity in angry but empowered-to-act student volunteers with PFC activity in students who believed they were helpless against their tormentors. Would the impotent condition generate more withdrawal emotions, such as depression, than experienced by those helpless to lessen their afflictions? (A 2004 New Yorker article, for instance, reported pervasive depression among young Saudis, for whom there are no jobs and no future opportunities. In animal studies a similar condition is called "learned helplessness.")

Harmon-Jones's team devised an anger-inciting scenario.[24] The students learned they would get a 10 percent tuition increase for the next semester. Half the participants were led to believe they had no recourse, but must pay the increase. The other half were told a petition would be circulated in an attempt to prevent it. The students' EEGs were measured as they listened to a radio editorial announcing the tuition increase. The students also rated the extent of their anger. In students who believed protest could halt the tuition increase, the left PFC fired more intensely than in those who thought they were helpless. Those students who reported feeling the angriest were also those who were most eager to revolt. By bestowing upon a person the energy, resolve, and focus needed to transform an untenable situation into something positive, anger fuels the engine of change.

Anger, the positive, left PFC activation, may be a source of active will to remedy social injustices. Harmon-Jones quotes Malcolm X as saying that "they called me 'the angriest Negro in America.' I wouldn't deny that charge. . . . I believe in anger." Imaging studies have measured anger in the more cognitive process of assessing "fairness." One recent test involving players deciding whether to accept another's offer of money found that unfair offers evoked extra activity in brain regions linked to anger and disgust; the more active those areas were, the more likely the player was to refuse the offer.

Harmon-Jones is also looking at the neural bases of cognitive dissonance, how we process "counter-attitudinal information" that inflames our minds—such as the flood of boldface lies spewed by politicians in attempts to coerce. His model predicts that left PFC activity increases undertakings that serve to reduce one's cognitive dissonance. Context is a huge factor in PFC operations. Anger may be more than vexation or simmering boil, or less than rage. It is usually blended with hate, disgust, despair, shame, fear, and so on. Each variant of anger, says Harmon-Jones, may bear its own signature bodily and cortical response pattern.

Do primary emotions have distinct prefrontal pathways? Mario Beauregard wanted to scan people who could tap deeply into powerful, primal joy, fear, anger, and sadness. So he put out a casting call to the Method actors of Montreal's Academy of Dramatic Arts. For Beauregard, acting theory boils down to two different schools: one in which the actor simulates the outer appearance of an emotion; the other, the Stanislavski method, where the actor embodies a character's feelings by drawing from personal memories. This emotional selection process, such as Brando learned at the Actors Studio, enables the actor to create a dynamic, affective bond with the character. "By using circuit memory to recreate the saddest events in their lives, these Method actors," notes Beauregard, "are allowing themselves be invaded by sadness to the point they are pouring out real tears. That was our rationale: to study an expert group of emoters."

Beauregard had previously scanned the brains of nonactors in the throes of emotion. But in the thespians he saw a far more intense bonfire of neural activity. "The spatial extent of the activity was huge—it captured much more brain area," he says, fairly shouting with pride and joy himself.[25] Relative to emotionally neutral states, opposing emotions, such as happy and sad, activated similar prefrontal regions—OFC and ventrolateral and medial PFC—as well as the memory areas of the temporal cortex. "We had thought sadness and joy would be treated in very different ways," he adds, "but fMRI imagery showed that the same cerebral circuits treat these contradictory emotions."

It seems likely, however, that happiness and sadness correlate with distinct subsectors within similar brain regions. The neural architecture could resemble the layout of motor and sensory cortices, where sectors designated for index fingers are next to those for thumbs, or cheek near tongue. Beauregard hardly claims there are no differences in brain processing when we laugh at comedy or cry over a lost love. "We can't touch them using MRI," he says, "but if we use endocrine or neurochemical methods, we might see differences."

Happiness and Sorrow

Happiness research is the latest subfield in affective neuroscience, spawning such publications as the *Journal of Happiness Studies*, featuring articles such as "Personality, Psychosocial Variables, and Life Satisfaction of Chinese Gay Men in Hong Kong." (They're moderately happy.) How do you quantify well-being? Is it a sum of such measures as whether you've got the career of your choice, the spouse of your dreams, or enough money? Nobel Prize–winning economist the ever-ingenious Daniel Kahneman argues that we've yet to devise a decent scientific measure of it. Among his other endeavors, he's creating a new field called hedonic psychology, the study of the general human sense of well-being.

Steady across hedonics research are data indicating that money cannot buy happiness. Since 1956, the percentage of Americans who say they're very happy—around 39 percent—has remained constant. Poverty exacerbates unhappiness, but increased affluence does not seem to contribute to happiness. Happiness in individuals, moreover, appears to be a fairly fixed dimension of personality style. Witness the tale of the lottery winner and the accidental paraplegic: one year after their life-changing events, both were equally happy. Why? Many researchers use the personality homeostasis argument—that regardless of how ecstatic or depressed they get, most people return to a hedonic set point. People will rationalize themselves into a happy state, find new ways to maintain their set point. That is perhaps why someone can look back on getting fired and say it turned out for the best. On the other hand, millions of dollars' worth of high-end toys will not add all that much to one's happiness levels—it is said. This adaptation to improved conditions is called the "hedonic treadmill."[26]

Some experts suspect that neurochemistry regulates the set

point via the dopamine reward system. Acting as the brain's accountant, dopamine plays a role in making you feel elated or dejected, depending on your evaluation of events. For instance, if you expect to place first in the 100-meter freestyle event, you might feel dejected if you come in second. But if you anticipated not placing at all, you might be overjoyed with a third. With dopamine, they say, it's all relative. The neurochemical accountant insists upon adjusting the dopamine output to your set point.

This may explain why some CEOs continue to claw and scramble after ever more obscene piles of loot: they crave a higher dopamine fix. Like other kinds of addicts, they're compelled to do whatever it takes to get it. Conversely, happiness and well-being may rev up the dorsolateral PFC, fueling its executive powers. Thus there may be a feedback loop in a person's life whereby more prefrontal dopamine leads to greater computational power, which leads to more brilliant ideas on how to invest capital, or design better software, which leads to more money, prestige, and more prefrontal fuel, and so on.[27]

Does the PFC process happiness differently than it does other emotions? A German team compared the neural networks involved with retrieving happy versus sad autobiographical memories. As a rule, personal memory is saturated with emotional content. Retrieving a long-stored memory, as Michael Petrides says, requires prefrontal operations. Volunteers were asked to summon happy and sad memories from their distant pasts, ranging from their earliest sad memory to their first pleasant memory of playing with others, first memory of love, first death of a loved one, and so on.

When people were scanned during these memory trips, the orbitofrontal cortex and its adjacent PFC and ACC neighbors in both hemispheres were major triggering stations for calling up the emotionally charged episodes. But within the OFC there was a clear distinction between zones engaged in retrieving happy and unhappy memories. Sad memories activated the lateral—outside edge—orbitofrontal cortices on both sides. Happy memories activated the medial areas (the ventromedial region, including BA 25) and other parts of the ACC. The patterns were distinct. This supports the notion that the orbitofrontal/ventromedial cortex, plus ACC, form the hub of an emotional-memory retrieval network—with "inner and outer" divisions for happy and sad memories.

Contrary to the belief that we recall traumatic events more vividly than happy ones, the subjects reported that their bucolic memories were more intense and were recalled with more vividness or "mental visibility." Sad and happy memories each lit up both hemispheres—contrary to the hemispheric lateralization hypothesis. This might happen because recollection calls for collaborative neural operations; or, since memories are composites of both verbal and spatial elements, both functions work together to assemble complete mental representations. That the rostromedial PFC (BA 10) also fired up supports observations that the interface between memory and emotion is essential to self-reflection, to creating and maintaining a sense of self.[28]

Regret: A Counterfactual Emotion

Classic gambling tests highlight another, subtle orbitofrontal operation: control of regret. Regret is a sour sentiment often attached to the admission that you made the wrong decision. Missed opportunities, the realization that we took the wrong road, may trigger regret. (Thus Yogi Berra's famous dictum: "If you come to a fork in the road, take it.") Regret's more bitter siblings include self-reproach, remorse, and shame. A test designed to elicit regret in gamblers when they chose a losing strategy compared the emotion in normal and OFC-damaged players. Facing the consequences of a choice can trigger emotions ranging from satisfaction and relief to regret, explained Nathalie Camille, the leader of the European study, reflecting our assessment of what we actually gained versus what we would have gained by making an alternate decision. In the experiment, game players with normal PFCs expressed regret when their gambling decisions cost them money. The OFC patients, however, showed no regret when they opted for losing moves. And they ended up with considerable net losses.[29]

When faced with mutually exclusive options, the choices we make are conditioned by what we hope to gain—the economist's "expected value"—but are also influenced by how we hope to feel afterward. We want to maximize our pleasures, minimize displeasures. Dread of regret, says Camille, profoundly impacts decision-making and is a powerful predictor of behavior, because our choices are often made to avoid this self-condemnatory emotion. PFC-mediated regret is triggered by our capacity to reason counterfactually, the "what if" cognitive mind-trick we saw reflective of general intelligence in the "Blame the Victim" study. Contrary to disappointment, which we feel when "stuff happens" independently of our will (by fate, accident, and so on), regret is powerfully associated with a sense of responsibility. It plays upon our sense of self as a free agent.

The healthy gamblers showed the usual bodily indicators of negative emotion when they realized they'd picked a losing strategy. But the orbitofrontal patients didn't. Seeming to experience no regret at missed opportunity for winning, they did not sweat more, their hearts did not race. This lack of bodily response at the visceral level confirmed the patients' lack of emotional reaction to the outcome of their bad strategizing. Nor was the patients' lack of regret due to lack of interest in winning. Indeed, when they saw piles of chips building up (or down) from one trial to the next, and kept track of their earnings, they responded emotionally to winning and losing. They just didn't feel the need to kick themselves after they'd made a bad bet.

This is evidence that emotions attached to decision-making are also attached to assessments of abstract, hypothetical outcomes of it. That is, when we are faced with choices, we pair the choice taken with options not taken in emotion-filled parallel universes of the mind. It is this counterfactual thinking—comparing the real to the virtual—that determines the quality and intensity of our emotional response. The absence of regret in orbitofrontal patients also suggests they have difficulty grasping the concept of personal liability for one's decisions that colors the emotion experienced by most of us.

Camille argues against the theory promoted by Damasio and others, who emphasize the "bottom-up" visceral-autonomic and subcortical influences of emotions on PFC decision-making. She proposes that the orbitofrontal cortex exerts a top-down influence on emotions "as a result of counterfactual thinking, after a decision has been made and its consequences can be evaluated." This, they say, is good for survival in the real world. The powerful emotional tag affixed to your cognitive choice, manifested as your feeling of responsibility for it—here, regret—reinforces learning and decision-making processes, in a kind of cognitive-emotional-cognitive loop. This intertwining of cognition and emotion is something excitingly intrinsic to higher-level thinking.[30]

Liking and Wanting

Personal preference governs most choices we make. Although we often choose what we need, selections in food, friends, or fashion are not necessarily choices made at the behest of immediate rewards, and they often involve making subtle distinctions between equally viable options. Preference-making draws on different PFC processes.

In a study at Emory University by addiction experts, volunteers answered a questionnaire measuring the intensity of their preference for such items as fruits/vegetables, cars, sports, and hobbies. Activity in the subjects' rostromedial PFC (BA 10)—the front-and-center self-referential zone—correlated well with strongly liked items. In contrast, take-it-or-leave-it items activated subcortical reward-processing areas alone. We probably generate mental representations of these desired things in our self-monitoring PFC system, but don't bother to do that for more quotidian stuff.[31]

The idea that pleasure is a reward for acting in a survival-enhancing way—eating, mating, finding shelter—underlies most hedonic theories. The British guru of the orbitofrontal cortex Edmund Rolls proposes that emotions per se are states induced by rewards and punishments. A reward is anything man or beast will work to achieve; a punishment, anything he will work to avoid. The mental processing of pleasure is thus defined in Rollsian terms by the positive event that triggered it.

Today, pleasure—"reward" in its overtly human form—is viewed as a pivotal force in cognition, influencing decision-making at the highest levels. Pleasure may help us focus on goals and filter out the irrelevant. Indeed, one might imagine that even before fear, pleasure may have been the primal emotion. There, pleasure was the sole affect, the one that inflamed curiosity. In the Garden of Eden it was all BAS, all positive, approach vector; nothing aversive. There was no amygdala in Eden, and no conflicting drives. Back in the reality-based world, Michel Cabanac at Laval University in Quebec City thinks pleasure evolved solely to prioritize eating, mating, sleeping, actions in the face of conflicting drives. People, for instance, strike a balance between the pleasure of eating and the displeasure of spending money at the market or the restaurant.

Affective brain science, like other cognitive sciences, may require a new epistemology to explicate the functional landscape of pleasure and other emotions. When do we know pleasure, happiness, or anger? And is it still pleasure (or anxiety) if it is not attached to a palpable reward (or punishment)? Can one focus solely on conscious feeling as the defining feature of emotions? Probably not: all manner of measuring systems, such as David's Zald's irritable-people data, support the idea of implicit emotional states. The theory of emotional style is based on the notion that subcortical neural circuitry is weighted toward specific affective tendencies.

Psych lab experiments on subliminal advertising—co-opted as far back as the 1950s by Madison Avenue—present evidence for this unconscious emotional brain processing. In one classic study, a person is subliminally presented with happy smiling faces. The viewer may not consciously experience a warm glow, yet will nonetheless drink more of a branded fruit beverage than if the happy faces had not been flashed before his or her unseeing eyes. Conversely, if the viewer is subliminally presented with scowling, menacing faces, he may drink less of the beverage. Because this stuff works, it is being used nonstop in the pitching of consumer products and politicians. Remember the Republican strategy of subliminally presenting the word "rats" in an attack ad in the 2000 presidential race. And that's just one we know about.

We are just at the surface of understanding the neural substrates of pleasure. Kent Berridge and colleagues at the University of Michigan have proposed that positive emotion in the brain may in fact operate through two overlapping pathways: one for "liking/pleasure" and another for "wanting/desire." According to Berridge, ground zero for pleasure is a sub rosa "core liking" system, a "strong, pure thrill," as he puts it, that may lurk in subcortical limbic and brain-stem circuits. Although these processes do not generally make themselves directly known, they can instigate awareness of liking when their input reaches the orbitofrontal PFC and other cortical staging areas of consciousness. As for "wanting/desire," the other system in Berridge's model, it may not be a hedonistic affective process at all, but a kind of drive state.

Berridge's prototype for "liking" is the lip-smacking, grinning relish with which a baby responds to the taste of a sweet sugary liquid. The bitter, biting taste of quinine, on the other hand, elicits from the infant a negative grimace. Other mammals, as people with pets know, display similar facial reactions to these two opposing tastes. Berridge thinks liking/pleasure brain circuitry differs from that of wanting/desire, both in wiring and in the neurochemicals that course through it.

For decades, dopamine reigned as the king of reward. The highway of dopamine that stretches from the midbrain to the mesolimbic emotional areas has long been regarded as the yellow brick road of the brain's reward system, which in turn was considered synonymous with pleasure system. The consensus was that dopamine release in this circuitry fueled all reward, therefore all pleasure-seeking behavior, in every organism from mice to elephants. Multiple reasons strongly implied that dopamine neurons mediated pleasure. Dopamine neurons in rats are turned on by pleasurable events, such as eating delectable new foods or encountering a sex partner. In humans, as well, dopamine release is triggered by drugs sought by multitudes—amphetamines, heroin, cocaine, and Ecstasy.

But Berridge noticed holes in the dopamine theory. When he blocked the dopamine system in rats, yes, they stopped eating as predicted. But even with all dopamine spigots in the brain shut off, when the rats were force-fed sugary liquid, they still exhibited the primal "liking face"—suggesting they still derived some pleasure from the sweet taste. Berridge wondered if the dopamine system, rather than controlling a circuit generating liking and pleasure, instead mediates "wanting/desire." Others, too, suspected that this system might not produce pleasure per se but elicit positive urges: sensations such as eagerness to engage the world or a sense of power. Which may not be quite the same as pleasure.

A regime change was in the works, and dopamine was about to be deposed as pleasure's potentate. Pleasure, Berridge thinks, is fueled by another group of neurochemicals, the opioids: endorphins and enkephalins, whose receptors are widespread throughout the cortex. Berridge increasingly suspected that liking/pleasure is propelled by opioid-powered circuits that overlap with the dopamine system, but are distinct in their mission to mediate pleasure. Where are these circuits? Leaving the heightened realms of the PFC for a moment, we descend to the limbic area's nucleus accumbens (NAC). Noted for its processing of rewards, this region lies just below the PFC. Berridge proposes that the center for liking/pleasure is the "shell" of the NAC, which enfolds the NAC core like an "elongated pastry pie shell." This outer NAC region's activity directly causes increases in "liking" reactions to sweet tastes, and these activations involve a special opioid circuitry.[32]

The shell of the NAC "liking system" is part of a neural system for preference and pleasure that includes such subcortical structures as the ventral pallidum. Lying below and behind the NAC, the ventral pallidum receives "liking" signals from it and relays them to parts of the thalamus, which in turn channels them to the orbitofrontal PFC. The ventral pallidum, says Berridge, is "a central fulcrum" for the circuitry of liking/pleasure, as well as a jumping point to prefrontal and other cortical systems of conscious pleasure. Electrical stimulation near the ventral pallidum in humans can induce bizarre bouts of mania that can persist for days. PET studies report that sexual arousal and competitive urges correlate with firing in this area.[33]

Although wanting/desire can include sensations that objects, people, or places are more alluring and compelling to pursue, wanting/desire is not about pleasure. The mood state of someone whose wanting/desire system has been activated shows what Berridge calls "incentive salience." A person may suddenly perceive the world as motivationally more attractive—even if no real liking/pleasure is present. This perceptive brightening and revved-upness is slippery and hard to describe, so we fall back on saying, "I feel good!"

Driven by the dopamine system, the wanting/desire pathway has as its central hub a different part of the nucleus accumbens. In lab animals, if dopamine is blocked, rewards seem to lose their luster; rats no longer pursue treats. But no neurochemical manipulations Berridge's team could dream up to suppress dopamine altered the animals' positive "liking" of the rewards. What blocking dopamine did truly accomplish was to turn off the rats' willingness to work for rewards, whether they were drugs, sex, or food. The rats ceased to "care."

Evolutionarily, "wanting" could be a primal form of goad to egg us on in our pursuit of innate incentives (trudging forward in a snowstorm to get to food and shelter) that's later harnessed to serve learned pleasures and liking. Dopamine confers upon mental representations of rewards their powerful lure whenever symbols of these rewards arise, thus the power of advertisement. Turning on dopamine systems causes a beer or SUV ad to become momentarily more intensely attractive. Seeing the beer cue on TV, you suddenly "need" a Bud. Essentially "nonhedonic," wanting/desire is often tethered to liking and pleasure mechanisms—so we chase after the desired thing. When working in tandem with opioid pleasure circuits, the dopaminergic "wanting/desire" system can translate pleasurable feelings into action. The two systems have been shown to share numerous pathways. (Dopamine, as the competitive and "push on" transmitter, may be genetically elevated in classic workaholics obsessed with results. The Nobel laureate Andrew Schally, for instance, worked straight through Christmas Day to isolate the LHRH molecule before his competitor Roger Guillemin. Beethoven composed one opera with four different overtures. Different wanting/desiring systems?)

Wanting/desiring obviously can be a rational PFC operation—if it is proportional in intensity to liking. That is, we consciously and rationally "want" those things we expect to like. Sometimes, however, they are out of sync, "decoupled." Take the example of a cake in the bakery window, cites Berridge, or a tax cut offered by a political candidate. You feel the allure of the cake and buy it; you vote for the candidate. When you eat the cake or get the tax rebate, however, the pleasure may not be proportionate to the strength of the initial attraction. (You then experience buyer's remorse.) The imagined gratification of wanting has gained an independent status in the brain.

Other times you only like something and do not want it. Young heterosexual men viewing the faces of exceptionally good-looking men and women indicated, for example, that they "liked" the men but "wanted" the women. This independence of liking and wanting makes possible the pursuit of something wanted but disconnected from liking/pleasure; disconnected even from the rational expectation of its pleasure value and other aspects of rational desire. Wanting to get one's teeth cleaned comes to mind as a mildly "nonliking" mental representation.

When wanting outstrips liking, it can become an addiction—irrational, waxing and waning with triggering cues. For the shopping addict, for instance, perusing a catalog hawking Fendi furs will provoke her to pick up the phone and order a coat, even though she is repulsed by the idea of ranching minks and is planning to join PETA and boycott furs. You can want what you cognitively don't like. Irrational wanting may affect the nonaddicted as well but perhaps not on such a regular basis. Is this "hyperwanting" an explanation for people's clinging to a politician long after the reasons for their enthusiasm have ebbed? Can vivid cognitive fantasy images substitute for external cues, triggering spontaneous mesolimbic wanting activation and irrational choice? Watch TV ads after one in the morning and the answer is clear.

How do the two systems hook up to the PFC? Focusing his research on the subcortical, Berridge posits that the PFC may be participating in the basic core processes of liking or wanting—or both—in concert with the accumbens and other lower systems. Another possibility is that the PFC mainly channels core liking or wanting into higher-order systems related to awareness of pleasure and desire. "That would be an interesting thing," Berridge offers. "To take the basic core processes and read them into consciousness. This presumes the PFC is mediating consciousness."

A third idea is that the PFC, as executive, inhibits or enhances our pleasures and desires to square with our plans and goals. In this model, PFC commands the willful regulation of pleasurable reactions. Thus we dial up or down our degree of liking or pleasure. A fourth possibility, Berridge concludes, is that wanting has another dimension. Influenced by research by Cambridge University's Anthony Dickinson and UCLA's Bernard Balleine, Berridge is coming to believe that the basic wanting process is not what we often mean by a conscious goal. Distinct cognitive kinds of wanting may exist that differ from more primal dopamine-related cue, sensory, image-triggered wanting. This higher-order cognitive wanting would include forms of conscious, declarative expectation, he explains. "You know what you want when you walk to the refrigerator. You know what might be there, and you imagine the possibilities. Most of our daily wants are of this sort—where we are cognitively spinning out the future. And our want is actually for an imagined possible outcome."

Older, primitive, subcortical, more purely dopamine-driven wanting often operates in the face of conscious desires and goals. Addiction, says Berridge, may be the best example of this duality of wanting. "Addicts often say they would rather not do drugs anymore. They know the outcome is bad—the pleasure of the drug isn't worth the pain of everything else." Yet when he walks into the bar and someone clinks ice cubes or lights up a cigarette, the recovering alcoholic orders a drink. Or the recovering cocaine addict encounters a current user, and finds himself unable at that moment to stop himself from using again. These are cases where whatever "want" they may have is not cognitively based.

Cue-driven wanting (seeing someone light a cigarette), Berridge suspects, seems to be centered in the accumbens, whereas cognitive wanting is much more likely to be PFC-based. "This possibility presupposes the PFC is doing something psychologically different from our lower-level cue-triggered wanting," he adds. "We don't know which of these possibilities is true. Maybe more than one of them is." Berridge hopes to explore the brain bases of more abstract forms of positive emotions: social joy, love, intellectual pleasures, aesthetic and moral appreciation. Do such complex positive, or approach, emotions share neural underpinnings with sensory "liking"? That, too, remains to be seen.

Where Reward Becomes Punishment

Until recently, the affective neurosciences studied negative emotion almost exclusively. The brain circuitry of aversity is thus far less of a mystery than the networks that generate hope and joy. Edmund Rolls suggests that the orbitofrontal cortex tracks changes in the reward significance of something if its "neuroeconomic" value shifts. Orbitofrontal neurons also track other switches in reward value, such as changes in the sensory pleasure of a stimulus. For example, monkey orbitofrontal neurons reduce firing in the presence of food when the monkey shifts from being hungry to being sated.

Few studies have tracked the neural dynamics of a human's mood swing from desiring and getting pleasure from something to feeling neutral to being repelled by it. One rare and cleverly torturous study involved a favorite comestible. Dana Small and colleagues at McGill University PET-scanned "chocolate lovers verging on chocoholics" while they ate chocolates beyond the point of satiety. The experiment, which took place close to lunchtime, involved having the subjects melt squares of Lindt chocolate (milk or bittersweet) in their mouths for a few seconds prior to being scanned. At this point, the person would indicate how pleasant or unpleasant the experience was, and how much he or she wanted or did not want another piece. The ratings ranged from "I really want another piece . . . delicious" to "Eating another piece would make me sick." The subjects were encouraged to eat chocolate to a punishing degree—up to seventy-four squares. Control subjects received a squirt of water into their mouths.[34]

When reveling in the chocolate, subjects showed activity in the lower medial orbitofrontal cortex, plus the insula and the striatum. The striatum facilitates the physical movements that initiate eating (I'm ready to pick up the food, etc.) and the underlies motivation to eat. The insula, the "gustatory" or "ingestive" cortex, serves to synthesize overlapping sensations from the stomach and other organs with emotional responses to the taste of what is consumed. The lower medial OFC network, however, blazed away only when the chocolate eaters were still jonesing for the candy. As motivation to eat ebbed, so did OFC activity.

When the sated subjects forced themselves to eat chocolate, and the reward value of the candy had long ebbed, blood flow in the medial OFC also decreased. Then another network involving the lateral OFC, left dorsolateral PFC, and right posterior cingulate became active. The posterior cingulate and the limbic parahippocampal gyrus serve to assay the emotional content of body states: "How do I feel about getting full now?" Since the dorsolateral PFC participates in the inhibition of inappropriate behaviors, this region, when receiving signals from other parts of the network to stop eating, could "decide" to override these signals in order to fulfill the terms of the experiment. ("I feel so stuffed I want to throw up, but I agreed to eat chocolate until this experiment is over.") Imagine the executive control Morgan Spurlock, the director/star of the documentary *Super Size Me*, must have deployed to choke down all those McDonald's meals.

As with positive and negative memory recall, here again are the brain's yin/yang functional patterns. As eating chocolate shifted from a rewarding to a punishing event, activity ebbed in the medial OFC and flowed to the lateral OFC. Lateral OFC activity may reflect our evaluation of a punishing outcome, similar to the way this region rates sad memories while medial OFC activity may reflect assessment of a rewarding outcome, operations it also performs for happy memories. Thus neural representations of reward and punishment may be segregated within these regions. The medial OFC may also be involved in monitoring and holding reward values online, while the lateral OFC comes online when an experience previously linked to reward must now be decoupled from it. This suggests there may be separate but interlocking PFC systems representing reward, punishment, and motivation.

When the medial OFC was buzzing, and the subjects said that eating chocolate was pleasurable, their behavior was in accord with their will. As their desire to eat faded and eating became inconsistent with their will, activity shifted to the lateral OFC. Thus lateral OFC activity began when the desire to stop eating was suppressed in order to conform to the demands of the experiment. The subjects' desire for more chocolate waned faster and more intensely than their reports of how pleasant or unpleasant was chocolate. That is, they didn't "want" it before they stopped "liking" it.

Love Is the Drug

The main thing to know about PFC ops during the hot stages of early love is that the executive branch of the brain is virtually offline, overwhelmed by a kind of "emotional flooding." The brain action in romantic love, according to a study by Andreas Bartels and Semir Zeki, mainly concerns four small brain areas, only one of which, a section of the ACC, is remotely higher-order. The British neuroscientists scanned a group of heterosexual male and female volunteers who professed to be head over heels in love, and whose statements were backed up by psychological and skin conductance tests.[35]

When the subjects gazed at photos of their sweethearts, different brain regions lit up than when they viewed photos of platonic friends. Besides parts of the ACC associated with euphoria, the loves zones included the middle insula, that "gut feeling" area, and parts of the putamen, the caudate, and the nucleus accumbens's reward centers. These regions also fire up under cocaine-induced euphoria. Besides the PFC being out to lunch, activity also ebbed in other "alertness" areas, notably the usually vigilant amygdala. This shutdown of the PFC in amorous states could explain the prevalence of "love is blind" clichés, and why lovesick people can make poor judgments or bizarre choices.

Critics of this study abound, however. The images are clear, but the emotions aren't, remarked Marcus Raichle of Washington University, a pioneering neuroimager. Although there seems a common reaction, what is the "state" being elicited? Are Bartels and Zeki studying what they think they're studying? asks Simon Le Vay, a neuroscientist with a Web site, Nerve.com. Bartels and Zeki's subjects all, reportedly, were in satisfying, ongoing relationships, suggesting that they were happily in love. But romantic love, notes Le Vay, is often unrequited, and decidedly less than ecstatic experiences can lead to depression, even suicide. "I'd guess that if Bartels and Zeki had tested a bunch of unhappy—but definitely 'in love'—lovers," Le Vay goes on, "they would have gotten a very different pattern of activity from the one they report. I question whether all the brain centers that lit up in their experiments are truly involved in the experience of love itself, rather than the euphoria of having one's love reciprocated, a euphoria that might apply to other emotional circumstances having nothing to do with romance." Indeed, passionate love may not be a distinct emotion but a kind of perfect storm of more fundamental emotions, such as wanting, sexual liking, and maybe even anxiety, reflected in patterns of brain activity.

That dopamine sinkholes of the brain were associated with the infatuation stage of love makes sense. Dopamine may provide the spark, causing people to feel a kind of cocainelike "wanting" high. It's unclear if couples whose love has evolved to the "next stage" would have the same brain scans as the impassioned ones. Another researcher reported after studying three hundred couples that amatory ardor usually lasts no more than two or three years. Cindy Hazen, at Cornell University, speculated that neurotransmitters associated with hot-blooded romance have a diminishing effect on brain tissue over time, in the same way as drug tolerance builds with time. When the heat fades, people often think something is wrong with the relationship, she said. But based on neurochemistry, this is to be expected.[36]

In 2004, Bartels and Zeki followed up their love studies, comparing maternal love with sexual passion. Identical brain areas were engaged when new mothers gazed at their babies as when lovers eyed images of their beloved. "Both are linked to the perpetuation of the species, and therefore have a closely linked biological function of crucial evolutionary importance," states Bartels. And both turned off prefrontal and related regions. Bartels and Zeki conclude that "human attachment employs a push-pull mechanism that overcomes social distance by deactivating networks used for critical social assessment and negative emotions, while it bonds individuals through the involvement of the reward circuitry, explaining the power of love to motivate and exhilarate."[37] (Meanwhile, little research on the neural bases of hate per se is available, but one suspects that its variations—hostility, enmity, antagonism, rancor, and so on—are rooted in subcortical networks akin to the primal fear pathways, with little PFC input.)

Emotional Regulation: You Can Change Your Life!

If you were born a sad sack, phobic, or necrophiliac squatting Diogenes-like in a garbage can, despair not. It may require some mental sweat, drugs, or both, but thanks to the PFC, you just might rewire some circuits, rejigger some neurochemicals, nudge that emotional set point a few degrees further along the BAS peace-and-happiness scale, and climb out of the Dumpster.

As a personal manifesto, Mario Beauregard believes that willful self-control represents one of the most potent mental facilities to emerge during human evolution, and has edited a recent book on the subject, *Consciousness, Emotional Self-Regulation and the Brain*. Beauregard wanted to reveal the neural substrates of volitional mood control, to unmask the dynamic neurochemistry of conscious, willed emotional regulation—specifically that of serotonin. So he called the Montreal actors for an encore.

The ubiquitous serotonin is studied primarily for its role in depression and other mood disorders; indeed, selective serotonin reuptake inhibitors (SSRIs) such as Prozac and Zoloft rub shoulders with aspirin and Advil in America's medicine cabinets. But there is far less scrutiny of how serotonin modulates normal emotional processing. Beauregard thus asked the actors to again induce in themselves states of intense joy and sorrow while his team tracked changes in their serotonin metabolism. This time he employed PET technology to measure how serotonin—5-HT (5-hydroxytryptamine), as it is known in biochemical circles—fluctuated as the actors experienced extremes of emotion.[38]

The actors set out in a neutral state of mind, but within fifteen to twenty minutes after they began to concentrate on personal imagery that evoked happiness, a tidal wave of change in 5-HT levels swept through their orbitofrontal PFCs, as well posterior cortical areas involved in visual imagery and the brain-stem wellsprings of serotonin. It happened by the sheer force of the actors' will-to-happiness. Conversely, as they channeled the sources of deep sorrow, another dynamic shift of 5-HT activity occurred in the PFC, visual cortex, and subcortical structures. But patterns of serotonin metabolism were not consistent within the same areas, and there was much more 5-HT activity in orbitofrontal areas during sadness than in the happiness state.

During both intense states, serotonin activity simultaneously increased in the anterior PFC region (BA 10) involved in self-reflection and analysis of one's emotional states, as well as visceral brain sectors that regulate the body's emotional sensations. These changes were wrought in top-down fashion, starting with the PFC initiating a cascade of metabolic changes through the brain. Since actors are professional emoters, Beauregard noted, they may be more proficient at voluntarily summoning peak emotional states. But we are all actors in our own lives, and our prefrontal systems may be able to elicit patterns of serotonin metabolism of our choosing. In the future, intentionally evoked mood states that induce these neurochemical cascades—based perhaps on meditation, biofeedback, and self-hypnosis techniques—may be among the most productive methods for treating mood disorders and traumatic mental conditions.

Can free will control the emotional brain? Some of the highest marks on the positive-emotion BAS scale go to Tibetan monks. The world's most accomplished meditators, they spend much of their lives contemplating compassion, controlling their anger, striving to banish negative BIS-type affect. During some EEG studies of emotional asymmetry, participating Buddhist monks show virtually no activity on the putative "dark" right hemisphere. Presumably a monk is not born this way: this is a happiness born of discipline. The practice of meditation and all that goes with it—or is suppressed by it—may have altered the monks' neurometabolic patterns.

When Richard Davidson first tested his frontal lobe asymmetry theory on a Tibetan lama in the early 1990s, he was astounded to find that the monk had the most positive, left-PFC-dominant emotional valence he had seen among the 175 people he had by then hooked up to EEG machines. Since the findings suggested that meditators might be inducing their own positive affective states, Davidson attached numbers of meditating monks to EEG machines. His recordings confirmed that their brain activity is qualitatively different than nonmeditators—down to the most basic autonomic twitches. One can literally explode a firecracker behind a Buddhist monk and he won't exhibit the so-called startle response. This near-universal reflexive measure of anxiety and uptightness—emanating from brain-stem circuitry—is generally involuntary. But with these monks, there is no eye blink, no mouth quiver, no heart-rate response.[39]

In 1997 Davidson took his meditation experiments a step further—to high-stressed Americans—when he enlisted Jon Kabat-Zinn, the founder of the Wellfulness-Based Stress Reduction Clinic at the University of Massachusetts Medical School, to teach employees at Promega, a Wisconsin-based biotech company, a meditation method. According to the Dalai Lama, this practice serves to induce "a state of alertness in which the mind does not get caught up in thoughts or sensations, but lets them come and go, much like watching a river flow by."[40] At the outset, Davidson tested the volunteer employees on EEG and charted their emotional personality profiles. A control group of Promega staff members was also tested before the experiment began. Their reward for participating was to receive meditation training later if they wanted it. Kabat-Zinn instructed the subjects three hours a week for two months on meditative techniques aimed at directing, focusing, and sustaining attention.

Before the study began, Promega participants typically scored high on the negative, BIS side of the affective scale. After eight weeks of meditation, their emotional set points had shifted to the positive, Davidson reported, claiming to find significant increases in EEG activity in several areas of the "happy" left PFC that persisted for around four months after the experiment ended. The meditators themselves avowed they felt less irritated, more upbeat, energized, and involved in their work.[41]

The study results also supported the notion that a sustained good mood has a salutary effect, perhaps enhancing immune system function. Among the Promega staff, the more negative, right-PFC-tending individuals were less able to fight off colds and other immune challenges. Davidson injected both the meditators and employees who were not meditating with flu vaccine. Within weeks he reported that meditating Promegans generated more circulating flu antibodies in their blood than did nonmeditators—suggesting that meditation promoted an enhanced immune response to the virus. How a person's mood or temperament affects his or her susceptibility to infection and disease remains mysterious. Does the PFC network communicate neurochemically with specific organs and cells of the immune system? That is, could anxiety and stress, which alter neurotransmitter function in the right PFC, then alter neuroimmune communication? As Helen Barbas's studies show, the PFC is wired into key parts of the hypothalamus that serve as pivotal regulators of both neuronal and hormonal aspects of a person's overall stress-response system. So yes, it's possible. Other mainstream neuroscientists are conducting more detailed studies of prefrontal activity in meditators to see if they can replicate Davidson's results. Anterior cingulate explorer Jon Cohen is intrigued by reports that proficient meditators can sustain focus for abnormally long periods. Most people have a limited capacity for intense, prolonged mental attention and control, and find it stressful to attempt.

Fear Regulation

Today, neuroimages of fear are commonplace. Perhaps as an aftermath of terrorist attacks, or simply because fear studies were bearing fruit, 2003 yielded a cornucopia of new findings. For decades, it was known that fear conditioning—learning to be afraid of something—sears into the amygdala a memory of the terror-inducing stimulus and its response. Say a real or imagined "claustrophobia program" encodes in your brain an application for Edgar Allan Poe–scale freakouts in small enclosed spaces. Then how do you extinguish or decode this fear of premature burial, so that riding in an elevator, or sliding into a coffinlike MRI tube, for that matter, isn't a waking nightmare?

"The reduction of fear is an active, not passive process," declares Gregory Quirk, of the Ponce School of Medicine in Puerto Rico. Quirk points to the medial PFC as the site of extinction control mechanisms. With multiple neural pathways to the amygdala and the autonomic centers, the medial PFC, including parts of the ACC, can effectively reach down and shut off the fear response. Quirk thinks an "all's clear" signal forms in the amygdala, close to neurons where the original fear memory is encoded. Damage to the medial PFC doesn't prevent the creation of an extinction code per se, but injury limits its life span, since the damaged PFC system may no longer encode and thus "remind" the subcortical alarm centers that there's nothing to fear anymore.

When subcortical areas sense potentially scary conditions (the tiny elevator), the medial PFC assesses the situation, and if it seems okay (you're only going up two floors), sends the "code green" signal below, muting the alarm signals. Although a fearful memory may still be stored down there, the medial PFC prevents it from triggering anxiety. If someone has a defective medial PFC, however, he or she might not be able to extinguish the fearful response, and the terror of the enclosed space will return again and again. Some experts speculate that targeting parts of the PFC in anxiety-disorder patients, using an experimental technique called transcranial magnetic stimulation, might help them control fear. This notion inspired a number of parodies, including an offer from the Halfbakery Web site for a Fear Control Helmet to wear during those excruciating moments before you attempt skydiving. "Now, with safety helmets that magnetically stimulate your prefrontal cortex, Fear of Extreme Sport can be a thing of the past! An easily accessible dial lets YOU choose the level of fear you experience. It may be Stupid, it may be Dangerous, but now it doesn't have to be Scary as well."[42]

Seriously, we should honor phobics for their heroic contributions in the scanning tubes and elsewhere. From myriad imaging studies, it's clear that not only do phobics' subcortical fear networks run in overdrive, more extensive brain areas are also sucked into their fear processing than in nonphobic brains. Phobics' "all's clear" fear-suppressing PFC systems may not be firing on all cylinders. People with special phobias may have less action in the PFC than nonphobics when looking at pictures of the terrifying object. Is loss of prefrontal inhibitory control specific to a particular phobia? In Sweden, an Uppsala University team scanned people extremely afraid of snakes but not spiders, and others with intense fear of spiders but not snakes, while each group stared alternatively at snake and spider photos. And yes, the subjects with arachnophobia did not show decreased prefrontal activity when looking at snakes, while the PFCs of subjects with snakes phobias did not experience PFC slow-

downs when looking at lurid spider pictures. The Swedes concluded that effortful emotional suppression is associated with an
engaged PFC. And reduced PFC firing when a person is provoked
to fear is consistent with a loss of emotional control, and may reflect
the excessive fearfulness of phobias.[43]

Ouch!

In pain research, experiments are painful. Kenneth Casey at the
University of Michigan has long studied pain perception and control, and inflicted some hurt along the way. In one study, he PET-
scanned volunteers as a hot-chili-pepper-like lotion was applied to
their arms. When they felt the burn, a neural "pain matrix" consist-
ing of a vast subcortical circuitry, the ACC, and the lateral PFC
blazed in their brains. The degree of hot pain the men reported
and how much their PFCs lit up strongly correlated. Those with the
highest PFC activity were men who experienced their pain to a
lesser degree than others, and in these men the subcortical pain
network too was quieter.

But it was more complicated than that: the left dorsolateral PFC
seemed to block the pain sensors in one part of the pain network,
while the right PFC inhibited another sector of the pain network.
According to Casey, increased left PFC firing works to dampen the
unpleasant emotional effects generated by nodes of the pain matrix
in the midbrain and thalamus. Right dorsolateral PFC activity,
meanwhile, seems to weaken firing in the pain networks in the
insula, thus diminishing both bodily pain intensity and vexing
thoughts about it. Thus both PFC hemispheres work together to
mitigate the pain experience by dual pathways.[44]

Other studies affirm what some physicians, witch doctors, faith
healers, and quacks have always known: that people can truly get
pain relief from merely believing they are receiving it. The vaunted
placebo effect owes its quite real power largely to the PFC's top-
down regulatory powers. The very expectation of pain relief—
a prefrontal function—transforms pain processing in the lower
brain pain matrix. Demonstrating this phenomenon, a Michigan
team scanned volunteers who had been told they would sample a
pain-relieving skin cream the investigators knew contained nary a
molecule of analgesic. First the subjects were given electric shocks to

their "untreated" wrists while their brains were scanned. Afterward they reported the intensity of their pain. Then the fake analgesic balm was applied, followed by more shocks. And voilà, participants said they felt less pain than they did when they were shocked without the "salve."

The scans, likewise, showed weaker firing in parts of the pain circuitry after the subjects got the ersatz treatment. After the placebo's application, scans showed heightened firing in the dorsolateral and orbitofrontal PFC and ACC as well. Prefrontal regions fired more as participants anticipated receiving the shocks during the application of the fake cream than they were when they braced themselves for unmitigated pain. Buzzing in expectation of pain relief, the PFC actually instigated a reduction in lower brain region pain-matrix firing. Top mind over lower mind. The researchers, led by Tor Wager, suspect the PFC's anticipation activates a release of pain-relieving opiates cascading through the pain networks.[45]

Conversely, the flip side of placebos, "nocebos," falsely promise to inflict pain. With nothing painful in them, sham pain inducers nonetheless may elicit noxious and all too real symptoms like headaches and nausea, and do so by means of prefrontal operations. In one study, volunteers given nocebo pills were led to believe they would experience nasty side effects from the "medication," and indeed they did develop them. Here the PFC, anticipating a heightened pain, may prime the lower brain and body areas to prepare for it. Placebo or nocebo, pain is in part a high-level mental anticipatory representation of harm (sometimes in the form of dread); an evaluation, at least, of harm's potential for hurting and for negative emotions in the future. The PFC's control operations play a role in this neural representation of pain, both cognitively in its degree of severity and affectively in its emotional regulation. Both placebo and nocebo experiments emphasize how powerfully our PFC-driven expectations, emotional states, and beliefs can temper basic, externally derived sensory experiences—even as these attitudes and beliefs also bias our rational thought.

Prefrontal Spin Doctor

Willfully locking down your emotions can cost you physiologically. You witness a traumatic car wreck; embarrass yourself at work;

force yourself to swallow your anger; or prevaricate about your feelings to a close friend. While you can succeed in suppressing fear, masking disgust, dampening shame, or lying about how you really feel, you may pay a price in increased blood pressure and heart rate, muscle tension, stomach upset, and so on. There is, however, a form of emotional control that works without exacting a toll on the body. This "reappraisal" involves cognitively reframing a negative situation into a more positive representation. In such cases we tell ourselves, "Okay, I'm facing something horrible, but if I think it through, it is not so bad."

Stanford's John Gabrieli and Kevin Ochsner investigated the neural bases of this PFC spin control. "Say I show people a video of a gruesome surgical amputation," Gabrieli begins. "At first everyone is grossed out and unhappy. But when the observers can tell themselves the doctors wouldn't be amputating if it weren't saving the patient's life, that the doctors are rescuing the patient . . . when people actively do this kind of mental reinterpretation, their autonomic systems do not go into overdrive."

Writing on their findings, Ochsner, now director of Social Cognitive Neuroscience at Columbia University, quotes Hamlet observing that "There is nothing either good or bad, but thinking makes it so." "Although Hamlet himself failed to capitalize on this insight," says Ochsner, "his message is clear: We can change the way we feel by changing the way we think, thereby lessening the emotional consequences of an otherwise distressing experience." Because reappraisal consciously transforms an emotion, it differs from just putting a lid on it. There is a neural sequence. First comes a fairly automatic process, whereby you feel the emotional blow, followed by your evaluation of its contextual meaning, then evaluation of an array of possible responses to it. Take a scene outside a church: a woman is crying on the sidewalk. Is she attending a funeral, or is something else going on?[46]

Before you are aware, the amygdala-OFC network goes to work, detecting the emotionally salient qualities in the church scene, setting it up for memory, and modulating the body's responses to it. The medial OFC evaluates the pleasant or unpleasant emotional weight of the event, ever sensitive to momentary changes in the event's social context and to your motivations concerning it. Ochsner suspects this amygdala-OFC circuit is in turn modulated by higher PFC operations when you reach the stage of emotional reappraisal. In reconfiguring the scene, you come to realize the woman crying in front of the church may actually be an effusive wedding guest, overcome by a flood of felicitous emotions.

The call for this shift in emotional context comes from the dorso-lateral PFC and its attendant executive processes. To see this net-work in action as someone intentionally transforms the emotional tenor of an affect-laden circumstance, the Gab Lab scanned young women volunteers. Like many other researchers, they used female subjects because women seem to respond more intensely to negative visual images. Some of the more gruesome images included a gun-shot wound, a charred body, and a blood- and vomit-spattered toi-let. The women were asked to invent a story about each photo to defuse its negative impact. With some practice, they successfully scaled back their distressful reactions to even the most lurid pictures.

The women's reappraisal of the disturbing scenes correlated with blazing left dorsolateral and rostromedial PFCs. Also active were the ACC and several posterior cortical areas. The more intensely these areas were engaged, the greater the morphing of unpleasant feeling into something less upsetting. And the more the PFC worked, the more the emotional linchpin regions of the amyg-dala and the medial OFC were put on hold. Those areas that had been aflame when the women first viewed the photos and regis-tered their emotional impact were now as quiescent as when the women gazed at photos of file cabinets.

Dorsolateral PFC neurons involved in reappraisal are those that also run cognitive control operations in working memory, maintain-ing information online while fending off distraction from compet-ing inputs. That both cognitive processing and reappraisal efforts arise from the same or adjacent dorsolateral systems suggests that an overlapping set of prefrontal ensembles handle cognitive regu-lation of both feeling and rational thought. It is not surprising that area 10, too, was also activated, since reappraisal of emotional sig-nificance implies its relevance to "self." ACC activity might again reflect this area's role in anticipating conflicts, such as between our first disturbing feelings and the rehabilitated more positive ones.

A key aspect of reappraisal may be this decisive shift of the PFC's center of focus from an emotion-laden process to a more dispassion-ate analytical mode. So how does the dorsolateral PFC diminish the screeching of the amygdala, turn down OFC activity? One notion holds that the dorsolateral PFC reaches downward through intermediaries in the posterior cortical regions. Perhaps the dorsolateral PFC operators dig into the inventory of perceptual memory banks throughout the cortex to do a makeover on the mental representation of the event. The woman crying outside the church is thus transformed from a grief-stricken figure into a nuptial celebrant in part by searching for and retrieving archival representations of weddings. At which point the amygdala alarm centers are commanded to register this glowing new mental narrative and basically shut up.

But why does the left brain dominate here? One reason, perhaps, is that verbalization—mentally talking to oneself—accompanying reappraisal may be an integral part of suppressing emotions. Ochsner hopes to conduct a study in which subjects will explicitly refrain from talking to themselves as a strategy, and instead employ more visualized, detached, third-person techniques that, in theory, should recruit right-hemispheric spatial processing systems. Or left-brain dominance in reappraisal might be part of the asymmetrical division of brain labor that tends to localize negative emotions in the right hemisphere and positive emotions in the left. Activity in the left PFC would reflect the engagement of systems supporting brighter feelings.[47]

Emotion and Rationality (the Yin and Yang Of)

Since the Greek philosophers, emotion and reason have been conscripted into a dance of opposition: one "loses his mind" in the fever of sick passion. Yet rational thinking "would be rudderless without emotion," declares Richard Davidson. And beyond frenzied extremes, reason and emotion do not truly stand in dialectical tension. Emotion is generally not a loss of self-control, but rather a powerful influence on reason, says Ray Dolan.[48] Nor is emotion a second-class citizen in the kingdom of thinking. Essential to the quality and range of everyday experience, emotions imbue events, ideas, and schemes with meaning. Emotions not only classify, rank, and codify the value of things but facilitate judgment about them. Enhanced memory for emotionally powerful events enables us to make better predictions regarding their reoccurrence.

Like a fixer in politics, emotions work to grease the skids. Emotional states support an "intellectual" bias to help resolve dilemmas ("I feel we should recheck the compass heading"), prioritize some cognitive tasks over others, and facilitate trade-offs. Emotions capture our attention, leading us to be better summoners of thoughts and plans. Until the "Affect Revolution" in neuroscience, however, topographers of the mind shunned the emotion-cognition nexus. Because feeling states are composed of nuance and shading, theories of emotion tended to implode in a vast vagueness. "People are finally starting to appreciate complex distinctions between different emotions and their connections to cognition," remarks SCAN Lab's Jeremy Gray. "For someone like me it's great. It means I have job security," he laughs. "Because we're not going to figure this out tomorrow."

For Gray, emotion and cognition are vast, integrated neural operations. Among the many diagrams and metaphors used to juggle the elements in this matrix, he features the analogy of a baseball team, seen with a touch of Zen. The baseball team is both separable from and interdependent upon its components. Pitchers must have catchers; separate positions, in essence they are defined by each other. Emotion and cognition are likewise players on the same team and defined by each other. Although these two systems of mental architecture may be largely anatomically segregated, they are also functionally, intimately intertwined. "Understanding both the team and game," Gray claims, "requires understanding each player not only in isolation, but in the context of the others."

To find out if emotional state influences the cognitive machinery of deliberation and goal-seeking, Gray devised a test involving emotion and working memory. After watching film clips intended to evoke one of two emotional states, Harvard undergrads played verbal and spatial computer games that taxed working memory. When the subjects viewed ten minutes of TV comedies from *Candid Camera* and a compilation from the *Best of America's Funniest Home Videos*, they reported being in an amused (approach) mood. When, while still amused, they turned to a challenging verbal working-memory task involving word recall, their upbeat emotional state enhanced their performance—versus when they performed the task in a neutral mood.

Amusement, however, actually impaired their performance of spatial working-memory tasks involving recalling a series of faces. But when the students viewed fear- and anxiety-inducing clips from the horror films *Halloween* or *Scream*, their performance on the spatial task actually improved. Less surprisingly, being scared impaired performance on the word task. Gray dubbed this dual effect "crossover integration." This was the first demonstration that specific moods have specific effects on "nonemotional" mental tasks. The more moved by the videos the students professed to be, and thus the more polarized their emotional state, the more powerful the crossover integration effect on performance.[49]

Each student had been "affective profiled," rated on the BAS/BIS scales that charted his or her sensitivity to reward-provoking or threatening situations. And they reacted to the videos according to these profiles. High-BAS people, for example, reacted more strongly to comedy. Each student's affective personality factored into how the video-induced emotion altered his or her working-memory performance. In keeping with previous studies of intelligence, students who struggled in the working-memory tasks also showed the most pronounced emotional biases. "High-trait" students—either extremely BAS- or BIS-oriented—were most polarized in their response to identical videos. On verbal working-memory tasks, BIS (withdrawal) personality types fared worse than BAS (approach) types across the board.

Collaborating with Todd Braver of Washington University, Gray next employed fMRI to pinpoint the locations of volunteers' brain activity while they were amused and made anxious via the comic or horror video clips, then asked to perform the same working-memory tasks. The findings were identical: word task scores were enhanced by lighthearted emotions and impaired by fearful ones, face tasks the reverse. Now, however, the experimenters saw where neural activity had shifted. The lingering emotional colorations altered firing patterns in the lateral PFC (BA 9) selectively in each hemisphere.[50]

There were the expected hemispheric specializations: words tended to activate the left PFC, faces the right. Students showed greater right-hemisphere firing during the pleasant mood condition when performing the facial recognition test, and greater left-hemisphere firing after being subjected to unpleasant emotional conditions during the verbal test. These combinations seem to create the most difficult mental climates, running counter to each PFC lobe's optimal operating conditions, thus making that PFC area work harder.

The lateral PFC, then, seems uniquely sensitive to the integration of emotion and cognition. Indeed, it was the only region to show this interaction, reflecting the psychological load of juggling both an intellectual task and an emotion—whether a helpful or obstructing one. "To our knowledge," Gray and Braver wrote, "this experiment is the most direct test of, and evidence for, the idea that PFC hemispheric asymmetries for cognition and emotion separately might mediate interactions." Are emotions and reasoning conjoined first in the lateral PFC? "The cognitive and emotional asymmetries are right there, like neighbors," responds Gray. "But that doesn't entirely 'prove' they're actually talking to each other. The integration could be taking place somewhere else, and then get projected up to these lateral PFC areas that are already sensitive to it. Some other, teeny tiny region we're not seeing could be causal. But I like to think the lateral PFC is where the action is."

What does this integration imply for our daily lives? Will you have optimal mental powers in your left PFC hemisphere if you are happy or carefree? And will your right PFC cylinders fire more efficiently if you're agitated or annoyed? If you're working on a problem that mismatches PFC activation with emotional state, should you expect to do worse than if mood and cognitive operation are in sync? Would watching a funny TV show for five minutes before work be more helpful if your job is predominately verbal—like presenting a legal argument in court? Would laughing-out-loud amusement impede complex abstract spatial thinking such as engineering design, math, or air traffic control operations?

"This PFC region cares about the pairing of an emotional state and task. You can't call this either an emotion or a cognition region. It cares about both what task you're doing and what mood you're in." Is this a conditional yes, then? "For optimal mental performance and the most efficient prefrontal network activity, you may need a match between the mood you're in and the type of mental operation you are doing," Gray grudgingly allows. "A good mood might help you in a verbal task, but if you try to get yourself in that mood, it could backfire, and you'd do worse than if you hadn't tried to alter your mood. So I'd be careful about making direct extrapolations or generalizing. But that's my job"—he laughs—"to be cautious."

"Of course, most mental challenges are not purely verbal or spatial," Todd Braver notes. "But if they are, yeah, to optimize your cognitive state—you should, at least in the short term, be in the right emotional mood." And what about personality types? Should a person opt for a career that matches his or her BAS/BIS profile? This question seems to freak Gray out. "I'd say no! I definitely would not advocate anybody choosing a career based solely on this information. It's not just the affective trait, but the interaction of the present state with your disposition."

But writers, a verbal bunch as a rule, are not necessarily known for their sunny, extroverted dispositions, I persist. "For the sake of argument," Braver considers, "let's say it is a positive personality style that makes someone a writer. But to write, you need the optimal level of arousal. Getting into the right mood state for writing may mean you need to increase some negative emotions because your negative baseline is lower than somebody else's. It's a complex pattern, to take into account the cognitive task a person is doing, the emotional state they're in, their baseline personality traits, as well as their baseline cognitive abilities." Is there is any purely neutral emotional state, idea, or memory? "Everything has emotional valence to it," responds Braver, "even though it may be minor relative to its other properties. I don't think there's a 'neutral' anything."

And finally, why would the human brain evolve such a braided and baroque system? Gray is toying with several theories. One is that it's an evolutionary coincidence, or "an uninteresting consequence of co-lateralization" of the hemispheres. "Yeah, that's quite a mouthful isn't it," he says ruefully. "It could just turn out that, well, hey, the PFC is a big place and the two functions, emotional and cognitive, just happened to be lateralized to different hemispheres, and it's more of an accidental result of the PFC's large size, rather than any functional reason." The "accidental tourist" theory of crossover integration.

Or, secondly, verbal and spatial tasks are behavioral "probes" for assaying brain function of more elementary prefrontal cognitive programs that verbal and spatial tasks tap into. And it's the computational properties or components underlying verbal or spatial processing that are really being pushed around by the emotions. The result is that an approach emotion, like happiness, enhances basic sequencing operations that were later translated into putting words in a right order, which is critical for the expression and communication of meaning. Or withdrawal emotions, like fear, enhance sustained attention in the right hemisphere, which then comes in handy for spatial computations such as rotating a three-dimensional object in your mind's eye. "Maybe the more elementary operations like sequencing or sustaining attention were co-opted by these emotions for our human efforts," says Gray. "It may not be about the words or faces per se."

And third is the neurochemical option: that dopamine and norepinephrine, for example, may "prefer" different hemispheres. Closely related molecules, dopamine and norepinehrine could be supporting slightly different computational functions in each hemisphere: dopamine for reward, norepinephrine for warning. Emotions could play upon these chemical systems to emphasize a specific processing pathway. And finally, what about a neo-Darwinian explanation, I ask. Verbal constructs, let's say, are inherently communicatory acts, and thus approach-related. But visual-spatial operations tend to be "looking out for," withdrawal mechanisms for vigilantly observing unusual patterns in the woods, scanning for predators, checking for escape routes in case you are attacked. Thus crossover integration is a brain architecture designed through natural selection. "It makes a nice evolutionary story, doesn't it?" Gray responds dryly. "But how do you test it?"

The Gray-Braver discoveries threaten to herald a new epoch in pop psychology, inciting a new wave of self-profiling questionnaires, self-help books, and offers to have one's emotional disposition calibrated, classified, and monitored along with your cat's BAS/BIS profile. "When the paper came out in 2002, we did get a number of requests from motivational speakers, from maximizing-your-brain-power types," Braver admits. "But I always emphasize that, first of all, the emotional states we induce are pretty extreme. If they were more moderate we might not have seen the crossover effect. We scanned people only five minutes after we induced the emotional state, so we don't know how long its effect on cognition lasts. There's more we need to know before one can, willy-nilly, give advice on how to maximize your emotional intelligence or personality."

In their scrutiny of individual personality types, Gray and Braver compared combinations of high and low BAS and BIS dimensions to assess the extent to which emotional traits contribute to and interact with reasoning intelligence. Their findings clearly indicate that high-BAS individuals perform better on working-memory tests, have better cognitive control, and thus higher general intelligence. Putting this together, a profile begins to emerge: a mentally advantaged individual shows a high degree of cognitive-emotional integration and tends to be essentially positive and extroverted. So then, are extroverts really smarter? Or conversely, does general intelligence predict for an outgoing, positive disposition?

The chicken-and-egg question is not solvable at this point, the scientists say. Some suspect that extroverts' greater social fluency and success may be the result of a more facile working-memory mechanism. That is, a high social IQ reflects superior executive processing in general, rather than special expertise or simply a greater desire to be liked. The extrovert, they say, excels at feats of multitasking: nonverbal decoding of others' emotional states from body language, tone of voice, and so on, while simultaneously tracking and carrying on conversations. Recall that Andrew Conway's cocktail party test showed the correlation between g intelligence and ability to screen out noise. (Some theorists of human evolution suspect that the social environment may have selected for bigger brains. That is, as the life of *Homo sapiens* grew more communally complex, the socially adept individual could better manipulate group situations to his or her advantage. Superior social intelligence, then, translated into better breeding opportunities—greater fitness.)

In the Gray-Braver study, low-BAS, high-BIS withdrawal personality types tended to do worse on demanding tests of PFC function. In cases where they performed comparably to the high-BAS types on working-memory tests, the anxious subjects exhibited higher anterior cingulate activity than the high-BAS subjects. Low-BAS people, it seemed, were compensating by stoking up their ACCs. In looking at their students' scans to see what this sector was up to, Gray and Braver found large variations in dorsal ACC activity. High-BAS people, who were more accurate on working-memory tasks, also showed relatively quiet dorsal ACC areas. The findings suggested that, again, personality differences in BAS and BIS dimensions correlate with differences in cognitive processing related to general intelligence, and that the cognitive dorsal ACC, within this network, plays a role in emotional personality and emotional state.

Interestingly, dorsal ACC activity did not seem to correlate with personality in the resting state—when subjects were disengaged from the challenging working-memory task. But as Zald has shown, the other, "emotional," subgenual ACC was very active during the resting state in anxious people, and so may serve to define a sort of "baseline" personality. That is, as one rather tightly wound researcher stated, "My baseline personality does not affect executive processing while I watch TV, but does when I play the piano or compete to run other drivers off the road." One might thus see how the ACC could play a continuous role in shading temperament, both under heavy thinking conditions or when a person is just chilling.

The dorsal ACC's involvement in conflict monitoring also may be at work here. Many rational thoughts involve emotional conflict. Even in a lab situation a participant may experience a conflict between the need to perform well on the task and the anxiety of failing to do so. And frequently we feel "conflicted" about selecting an answer—indeed, the larger the decision, the more likely the alternatives will include taxing reasoning and emotional conflict.

Is the ACC, then, the birthplace of personality? In modulating emotional states does it strengthen the scaffolding of temperament? Or does it help preserve an emotional equilibrium, so that someone less inclined to sunny moods may become more optimistic? Or conversely, does the ACC guide a starry-eyed optimist away from risky business? Gray suspects that the BAS component of the cognitive-emotional network will incorporate the dorsal ACC as a pivotal personality definer. But he adds, ACC firing here may merely reflect the activity of some other unit or network feedback loop sensitive to emotional states.

Big Brother

By charting the neural substrates of what turns us on and off, we are beginning to quantify the nature of human experience. Such knowledge, like most power, is a double-edged sword. It comes as no surprise that the field of neuromarketing is attempting to exploit the tools of cognitive neuroscience to "read brains," to seek out and channel people's inchoate passions and drives into windfall profits. What's troubling for the twenty-first century is that neuromarketers may be able to bypass conscious thought to tap directly into, say, liking and desiring systems to incite us to press the buy button.

Marketers have always sought to co-opt our reward-seeking, punishment-avoiding, stimulus-response, lever-pressing selves. But despite billions of dollars spent on motivation research, focus groups, and the like, human wanting/liking has remained so stubbornly opaque that as the Caltech economist Colin Camerer put it, "consumers are like some random finicky cat." Neuromarketing aspires to make that cat eat exactly the food it wants us to eat. Neuromarketing, adds Camerer, is rather like "interviewing the brain directly."[51] And the brain, supposedly, doesn't lie.

Articles in the popular press have depicted not-so-futuristic scenarios in which people strapped into scanners ogle potential new products. A piece in *Forbes* magazine, for example, asks, "Are the subjects really focusing on pitches for Kit Kat candy, Smirnoff vodka, and the Volkswagen Passat? Are they forming emotional attachments to these products?"[52] A sudden blaze of activity in the left orbitofrontal cortex: is this an "approach" response to the image of a Kit Kat chocolate bar? Does this neural excitement mean the subject is attracted to the Kit Kat brand image or message? Should the right OFC flare up, is it signaling subliminal withdrawal from the obnoxious, tongue-wagging character who pops up in a commercial for Carling beer?

At the behest of the marketplace, neuroimaging instrumentation is being used to explore the neurobiological basis of consumers' preferences and drives. In 2004, investigators at Emory University in collaboration with BrightHouse Neurostrategies Group, a division of BrightHouse, a so-called ideation consultancy based in Atlanta, scanned volunteers while they viewed diverse images—everything from a Ford truck to Coca-Cola, broccoli, Bill Clinton, a golden retriever, and Madonna. When the subjects eyed images they indicated they liked, the medial PFC—that preference zone—blazed. Other neuromarketing collaborations include the Mind of the Market Laboratory at Harvard Business School and Baylor in Texas.

Politics is another social application of this research. A 2004 study at UCLA compared the neural reactions of Democratic and Republican subjects to campaign advertisements. According to investigator Marco Iacobini, partisan brains showed partisan firing patterns when subjects watched a Bush commercial that made use of September 11 images and the infamous "Daisy" ad Lyndon Johnson wielded against Barry Goldwater in the 1964 campaign. The Democratic subjects' threat-sensitive amygdala was significantly more vigilant than the Republicans' when they viewed both spots.[53] At a subconscious level, then, were Republicans less bothered by what Democrats found alarming?

The investigators had two preliminary inferences for why this might be so. Democrats conceivably saw the 9/11 issue as a trump card in Bush's reelection strategy, and this hot-button imagery was threatening to them. But since the scientists noted the same activity spike in the amygdala when the Democrats viewed the "Daisy" commercial, with its juxtaposition of the daisy-picking little girl and the mushroom cloud, they also suspected that Democrats may be generally more alarmed by the use of force than Republicans.

While in the scanner, these same subjects viewed photos of presidential candidates Bush, Kerry, and Nader, as well as neutral pictures. The candidates' faces tended to excite the ventromedial PFCs in all subjects, regardless of party. But after looking at the Bush 9/11 segments, subjects then responded to the candidates along party lines. Their ventromedial areas still lit up, but when the subjects viewed the despised opponent candidate, their dorsolateral PFCs also blazed. Iacoboni speculates that the viewers identified with their candidate emotionally, but when they looked at the man from the opposing party they also deployed the more rational dorsolateral PFC machinery to marshal arguments against him.

Today, there are two ways of looking at the neuromarketing phenomenon in the political arena. In one, you need not time travel to a silicon-based dystopia to see how groups might exploit neuroemotional techniques to crack open the minds of the electorate and pry out affective data for partisan advantage. We know mind-shaping techniques are being practiced in politics' public relations wars, such as in the "framing" of issues techniques analyzed by the linguist George Lakoff and others. From Goel and Dolan's work, it is certainly apparent how influential and deeply embedded in logical thinking are irrational, emotion-dominant belief systems. These subterranean affective wellsprings undoubtedly can be tapped and harnessed to incalculable coercive control and profit. "People make tons of decisions and often they don't know why," Iacoboni said. "A lot of decision-making is unconscious, and brain imaging will be used in the near future to perceive and decide about politicians."[54]

Neuroimaging techniques now, on the other hand, expose almost nothing of the individual mind. Scanners show that a person is exercising a specific neural network, but don't begin to reveal the content or complexity of that person's thought. Thus it is easy for marketers to interpret the bright neon brainshots they scrutinize like chicken entrails according to their own wishes and aims. At present, says Jeremy Gray, marketers can do little with these things. "But we need to begin discussing the social and ethical implications now—before the technology becomes sophisticated enough to be effective."

4

VIOLENCE

Morality and the Minds
of the Killers

In Stanley Kubrick's shooting script for A *Clockwork Orange*, corrections officer P. R. Deltoid asks the rude boys: "What gets into you all? We've been studying it for damn well near a century, yes, but we get no further with our studies. You've got a good home here, good loving parents, you've got not too bad of a brain. Is it some devil that crawls inside of you?"[1] Similar questions are still being asked by cognitive brain scientists today.

Finding two people who suffered prefrontal injuries as babies was epochal to the Damasio lab. Since then, the group has added significantly to their roster of people with early childhood trauma to the brain regions of social intelligence. But in the mid-1990s, the appearance of "A" and "B" was akin to discovering rare samples of an exotic species. Neuroscientists now had the chance to observe nature's accidental "knockout" humans, not unlike knockout lab animals in whom specific biogenetic processes have been deliberately deleted. In this young man and woman, the circuitry underlying social awareness and emotional regulation never wired up at all.[2]

Brief portraits reveal how thoroughly injury-stunted "A" and "B" are in their ability to negotiate the social world. Both subjects were from stable homes with educated parents. Hit by a car when she was fifteen months old, "A" suffered extensive ventromedial and frontopolar damage. Although at first she appeared to recover, by age three "A" was beyond the reach of ordinary discipline. While she scored well on standard intelligence tests, her teen years were a maelstrom of abusive and violent confrontations and rule-flouting. She became a chronic liar and was arrested repeatedly for shoplifting and other petty crimes, indulged in risky sex, got pregnant at eighteen, and then neglected her baby. She showed no signs of empathy, remorse, or guilt. Unable to hold down a job, she fell back into a dependency on her parents—who were maxed out mentally and financially by this hellion in the midst of an otherwise normal middle-class family with well-adjusted siblings.

Subject "B" had a tumor excised from his right frontal lobe as an infant. Like "A," he'd made an excellent recovery and appeared to be developing normally. But by age nine, he was friendless, emotionless except for occasional explosive outbursts, and motivationless. He managed to graduate from high school, but without a job he lived in squalid living quarters watching TV, gorging on junk food, racking up debt, and engaging in theft and occasional assault. He, too, chronically lied and fathered a child whom he did not support. Like "A," he showed no signs of empathy, remorse, or guilt. Today, "B" is living in a protective facility and "doing relatively well," says Steve Anderson. And subject "A?" "Basically out and about—and a disaster."

The scientists were struck by how much worse the effects of prefrontal damage were in "A" and "B" than in adults with lesions in similar areas. Remarkable was the abject failure of their "socialization," the complete absence of empathy and moral reasoning. Stuck with the social reflexes and moral judgments of toddlers, "A" and "B" were generally just trying to avoid punishments. The rules of social life were nonexistent to them. While people who were injured as adults might act inappropriately in some circumstances, they still knew the social rules. "As long as we give them a nice verbally encapsulated version of a social or moral reasoning test, they do fine," says Anderson. "They can even give you good advice on your personal life, investments, things in the real world in which they themselves are a mess." The severity of "A's" and "B's" social clue-lessness suggests that there may be a finite neurobiological window of opportunity within which the brain structures that enable us to learn basic laws of conduct and how to relate to our fellows are formed. Finite windows of opportunity exist in some primary sensory brain systems: when the visual cortex is badly damaged early enough, for example, one is blind.

Anderson and colleagues are using their growing database of early PFC damage cases to investigate plasticity, or lack thereof, in the development of social cognition. "Take language," says Anderson. "A child can have a big stroke in the language structure that will leave him aphasic. But then he shows a recovery in language function that you or I as adults could never achieve. You can see amazing recoveries even with left hemispherectomy in children up to adolescence!" But Anderson is not seeing this same plasticity with social-emotional functioning. It's more like vision. We can't regrow prefrontal tissue, and nothing suggests that any other part of the brain can assume these functions.

What do "A" and "B" reveal about the normal neural underpinnings of social cognition, and the PFC's role in it? The Scottish psychologist Andrew Whiten and others propose that human mental acumen evolved in large part to facilitate the special complexities of our intensely social primate life. Social emotion, notes Ralph Adolphs, stands in a privileged position, "tightly coupled" to social cognition. This coupling is not only heavily controlled and regulated in adult social behavior but its basis is hardwired during infancy.[3]

Adolphs proposes a triad of neural systems operating in a hierarchy of social thinking and emotional processes. First, multimodal sensory brain areas compute the Other's physical states—appraising "body language," balance, position in space, and so on. Second, the emotional cabal of the amygdala, the striatum, and the orbitofrontal cortex correlates data about body states with emotion, motivation, thought, and action. Lastly, the PFC, with its neighbors in the parietal and cingulate cortices, constructs large-scale theaters of the mind in which you form dynamic representations of the Others, their relationships to you and each other, their goals in light of your goals, and the "value" of your actions in the context of this social environment. Considerable computing power is needed to build this evaluatory system.

A huge number of studies, starting a century ago and now in full flood thanks to imaging technology, involve showing volunteers pictures seeped in social significance, then observing what brain regions ignite as viewers react. Humans are such skilled readers of emotions, and so sensitive to the contexts in which they occur, that if a face signals the merest glint of fear, happiness, disgust, anger, sadness, we will call the emotion correctly. Indeed, the brain is such an efficient detector of others' emotions that merely the ghostlike cartoon pair of "wide-eyed" eye whites on a black background causes the "fear" neurons of the amygdala to spike. A distributed network, including the fusiform gyrus of the occipital lobe, and limbic areas of the temporal lobe, zaps around complex information conveyed by facial expressions, and in milliseconds we automatically register that, say, "She is happy to see me."

In real life we seldom see a face alone. We observe the whole person and react fast and powerfully to his or her body language. Moreover, in dance, theater, sports, countless rituals, games, and social events we take special pleasure in attending to the body signals of others' feeling and intent. Nouchine Hadjikhani, a Harvard Medical School radiologist, and her Dutch colleague Beatrice de Gelder explored neural responses to a fearful bodily expression. They videotaped actors performing emotion-fraught vignettes— being surprised by a burglar, or in fearful postures, with their arms and hands held up as if to ward off attack. They scientists pixilated out the actors' faces so as not to confound viewers' responses to body expression. As a neutral control scenario, the actors poured water into a glass.[4]

When scanned volunteers watched these tapes, their brains processed body information at least as quickly, and with the same structures, as facial expressions: the fusiform cortex and amygdala. Other researchers were surprised at Hadjikhani's findings, arguing that the subjects were actually mentally "filling in" the image of an unseen fearful face. So in a following experiment, de Gelder clearly showed that the firing of the fusiform cortex in response to the body poses occurred too quickly to be a consequence of fabricating an entire mental image of the face.

In fast cortical processing, much reading of others' body and emotional states goes on nonconsciously. An ingenious experiment on the neural algorithms encoding others' behavior underscores how similarly each of us experiences the Other. In this experiment, Uri Hasson and colleagues showed participants thirty minutes of Sergio Leone's *The Good, the Bad, and the Ugly* while the subjects were in the scanner. Using intricate data analysis programs, the Weizmann Institute neuroscientists saw a striking degree of concordance in the viewers' brain activity. Their brains, it seems, were collectively, stereotypically ticking in synchronized space-time patterns as the movie played out.[5]

The face-related fusiform gyrus, for instance, glowed when the viewers focused on close-ups of Clint Eastwood and other actors' mugs. Another region, the collateral sulcus, fired vigorously when they saw indoor and outdoor scenes, such as camera pans of landscapes, buildings, events taking place in a gun shop, or Main Street. This collateral sulcus, it seems, is a functionally specialized "circumscribed space" subunit of the visual cortex, located not far from the fusiform gyrus. When the watchers scrutinized close-ups of delicate hand movements, such as cocking a pistol, a higher-order motor area sensitive to movement of body parts and eye-gaze shifts was ablaze. "The collective correlation attests to the engaging power of the movie to evoke a remarkably similar activation across subjects," the scientists concluded.

This collective brain firing, they proposed, might be used as a baseline for exploring cultural differences among ethnic groups. Do Israelis' and Palestinians' brains light up for Clint Eastwood's hand movements in the same way? It would also be fascinating to know the impact of prior experience on a moviegoer's response—such as combat experience, or being an Eastwood fan and having watched this movie five times previously. These findings have philosophical implications, as well, tending to refute concepts that each person inhabits a solipsistic inner world not comparable to that of any other person; it further refutes the notion that there exists no unitary external world at all. During the movie, the neural traces in parts of one person's brain were very similar to those in others' brains while they viewed the same footage. But equally significant, there were brain regions that were not collectively in sync as the viewers watched the flick. These included parts of the parietal lobes and most of the prefrontal cortex. Thus, commented the Brown University psychologist Luiz Pessoa, there will be "ample cortex for you and me to experience *The Good, the Bad, and the Ugly* in unique ways."[6]

The British neuroscientist Tania Singer used pairs of romantic lovers to explore the brain bases of sensitivity to another's pain. With her male lover in the same room, a woman went into the scanner. Singer watched the woman's brain patterns as a one-second shock was applied either to the back of her hand or to her partner's. The man's face was hidden, but the woman could see which one of them was going to get zapped, and whether it would be a weak shock or a sharp jolt. As we've discussed, the brain contains two major pain pathways: one is primarily somatosensory, registering the origin and intensity of the pain; the second is more emotional, gauging how unpleasant you feel the pain. Thus how much the pain bothers you depends on the context and what else is on your mind.[7]

When the woman observed her lover getting shocked, the sensory map of her hand that her own pain activated failed to fire. Witnessing a lover in pain, however, automatically triggered her emotional pain circuitry, including the ACC and the anterior insula (that "me" area). This suggests that the empathetic experience of pain is rooted in discrete parts of the pain matrix, and that "mirroring" of the Other requires only these emotional inputs to generate the basis of empathy.

Singer's findings affirm the notion that the insula maps an image of one's body's internal state, which underlies awareness of the physical self as a feeling entity. The anterior insula fires more for the anticipation of pain, whereas pain's actual experience activates more posterior insular zones. This anticipatory anterior insula also fired when women "felt" their partners' pain. Indeed, insular involvement appears to be essential for neural processing of "self" and "other."

Singer thinks insular circuitry may serve to form dual mental representations of feelings: first, those enabling us to predict the effects of future pain or pleasure on ourselves; and second, those that allow us to predict how a particular bodily experience will affect another. The mental workshop model of the relevance of pain or pleasure to the Other that we create in this circuitry is independent of our own direct sensations of pain or pleasure. This decoupling may be necessary for us to "mentalize" the feelings, intentions, thoughts, beliefs, and fantasies of others.

Our brains may also mirror mental mistakes of others. People were scanned while they watched others playing a computer game.

A watcher's ACC—the conflict monitor—leapt into activity when he witnessed a player make mistakes. The level of brain activity was comparable to that of the person himself botching it. As well as demonstrating how we register forms of empathy, this phenomenon also points to ways we learn by observation: the ACC fires up when an event turns out worse than we expected. Of course the pattern of brain activation might be radically different if the two people were competitively pitted against each other, when one person's error becomes another's gain.

But beyond "infection," empathy involves the capacity to simultaneously sustain your own mental state along with the Other's, rather than just "catch" his mind-set and feel it as your own. More complicated forms of empathic judgment and deeper probings into the minds of others, as Adolphs notes, reflect "postperceptual" mental processing—the progressive decoupling of information from physical sensations, gestures, or facial expressions from that generated internally via memories, associations, and inferences. In this postperceptual realm, personality style, intelligence, and special talents play a bigger role in how one evaluates the minds of others using prefrontal skills.

We are not born with such talents. Indeed, research on social cognition in teenagers indicates that the PFC is not much help during adolescence. During puberty, cortical regions, especially the PFC, undergo extensive reconfiguration, and the turmoil within this construction site is reflected in how teens judge others. Robert McGivern at San Diego State University found that children's ability to read other people's emotions—in the sounds of words and facial expressions—actually declines as they move into adolescence. Teens, noted McGivern, have noisier brains than younger and older people, as the higher cortical areas undergo this final stage of remodeling. Thus it may be harder for adolescents to process information about emotional states of others. This may help to explain why teenagers tend to find life so unfair: they cannot read social situations efficiently and suffer for that—ironically, since peer acceptance is the epicenter of teen life.[8]

Hoping to tease apart the contributions of various PFC subcomponents to social-emotional thinking, Oxford's Edmund Rolls looked at thirty-five adults who had undergone precisely defined surgeries in either the orbitofrontal, the dorsolateral, or the medial

PFC areas. First, patients were asked to identify the emotion they detected when listening to various nonverbal human sounds—sighs, screams, grunts, sobs—as signifiers of others' emotional states. They were also to identify neutral voices and sounds such as a chair scraping across the floor, someone chomping into an apple, and the whir of an airplane engine. Second, they categorized the emotions in faces. Then they evaluated their own emotional lives since their surgeries. And finally, close friends or spouses assessed perceived changes in the patients' social responses—such as empathy, public behavior, and interpersonal relationships.[9]

All in all, Rolls found that orbitofrontal patients had not morphed into ultraviolent Droogies. Their problems were subtle. People with lesions in the OFC, in either one or both hemispheres, struggled hardest when inferring the emotional tone of vocal sounds. Even a small OFC injury was enough to make it hard for them to detect whether a voice was cheerful or morose. The same for people with ACC lesions. This, the scientist surmised, might be because a fiber tract conveying phonetic information very specifically links the temporal lobes to these OFC-ACC regions.

People with OFC damage—such as the Danish guitar player, LP—act in socially inappropriate ways in part because they cannot decode the emotions in people's voices around them. Imagine the problems in phone calls, business negotiations, or party conversation if you can't assess another's mood by voice. If we are in a theater audience, most of us can by listening to the actors' voices alone know what is transpiring emotionally in the play. Not so the OFC patient. Interestingly, these patients, who were severely impaired in identifying emotions in voices, fared as well as healthy people in identifying nonemotional sounds of the world around them. And they were much less impaired on facial evaluations. The distinctive visual association circuits in the OFC are more widely distributed than those for sound, so there may be more redundancy and resilience in the brain processors for facial expression.

In evaluating their own changed status, every patient with OFC lesions in both hemispheres professed to experience new difficulty in perceiving others' emotions, as did patients with ACC injuries in one hemisphere. But people with injuries to only one OFC hemisphere claimed their facility in evaluating the emotions of others had not changed. Yet they were among those who fared poorly on recognizing the emotional content of others' voices. Some patients with ACC damage, on the other hand, performed well on the voice test but declared themselves to be profoundly changed in their perceptions of others—so altered in some cases as to become "different people." They reported deteriorating relationships; they also felt they'd become more emotional, with escalating emotional outbursts, especially bouts of anger. Some felt themselves to be hypersensitive to sad events, getting upset far more easily and in one case, weeping more at the movies. One patient, on the other hand, reported that she now experienced enhanced pleasure in the world of nature, in music, and in friendships.

To family and friends, all the patients except for those with damage in only one OFC hemisphere seemed clearly changed. The biggest transformation to those with OFC lesions in both hemispheres was that they no longer seemed to notice when others were sad, happy, or disgusted, no longer comforted others who were melancholy or afraid, no longer seemed to care what others thought. They were less cooperative, more impatient and impulsive. They seemed more isolated and self-immersed. Patients with injuries to the dorsolateral PFC, or anywhere outside the OFC-ACC neighborhood, however, had no difficulties with any of these social-emotional evaluations.

Damage to the ventromedial PFC, that zone in the middle of the OFC, may impair the social-emotional evaluation of sarcasm. The Israeli psychologist Simone Shamay-Tsoory and colleagues compared patients with prefrontal and posterior lobe damage and healthy controls. All participants listened to brief audiotaped stories, some sarcastic, some neutral. Here is one sarcastic example: "Joe came to work, and instead of beginning to work, he sat down to rest. His boss noticed his behavior and said, 'Joe, don't work too hard.'" (Meaning: "You're a real slacker!") After each story, the researchers asked factual questions and attitude questions to check the listeners' comprehension of the speaker's true meaning: did the manager believe Joe was working hard? When participants got the attitude wrong, they got an "error" score in identifying sarcasm.

Participants with PFC damage were the only participants impaired in comprehending sarcasm. Within the prefrontal group, moreover, those with damage in the right ventromedial area had the most profoundly faulty sarcasm meters. Ventromedial injury will disrupt "getting" sarcasm because this attitude demands an understanding of social cues, empathy, and emotion recognition. Detecting sarcasm, and probably snarkiness in general, "requires both the ability to understand the speaker's belief about the listener's belief and the ability to identify emotions," the scientists say. The listener must grasp the speaker's intentions in the context of the situation. This calls for sophisticated social thinking and "Theory of Mind." When Theory of Mind is limited or missing, a person will have problems interpreting irony, the broader category of social communication into which sarcasm falls.[10]

The Minds of Others

In 1978 the psychologists David Premack and Guy Woodruff published a seminal paper introducing Theory of Mind (ToM)to define a set of skills enabling us to think about what others are thinking, feeling, or planning. Almost all social interactions involve these sophisticated inductive computations. Such is the nature of art and politics, business, movies, sporting contests, poker, chess, and hide-and-seek. This cognitive capacity may be unique to our species although other animals possibly have some version of it.[11] Inevitably, investigators have sought to locate a consistent set of brain areas that subserve it. Indeed, some evolutionary neuroscientists think that the high-level neural computations necessary for ToM drove the remarkable development of the human cortex and the PFC specifically.

The earliest manifestations of ToM emerge by age four. When children, starting to construct internal models of their environment, begin to realize that their own mental worlds may depart from reality, they also begin to realize that one person may think something different from another. One subset of ToM online by age four is the attribution of beliefs to other individuals, especially false beliefs. A classic test of false beliefs requires a child to track another person's state of mind. A youngster watches a cartoon character, Sally, drop her toy dog Fluffy in a basket then leave the room. Soon thereafter, Ann sneaks in, removes Fluffy from the basket, hides the toy, and exits. Sally returns and starts looking for Fluffy. The crux of the question is: where will Sally search? At four, children realize that Sally will mistakenly look for the toy in the basket where she left it. That is, children understand that another's mind may have other angles of perceptions than their own, even false ones. This is, incidentally, about the same time working memory starts to function with a greater degree of flexibility.

False beliefs and, more intensely, deliberate deception, are basic elements of life's drama. As a theater audience, we experience a thrill of omniscience as we eavesdrop on stage characters blundering ahead under their burden of delusions, of sometimes comic, sometimes tragically errant theories of mind. Part of Shakespeare's genius is his mastery of Theory of Mind. In *Othello*, for instance, tension is power-generated by the wheels-within-wheels of false belief. We attend to Iago as he parses the mind of Othello, spinning the Moor's pride and jealousy and warping his rationality by implanting the false belief that Desdemona is having an affair with Cassio. We witness Othello lured into acting upon this false belief about his wife's state of mind. Desdemona, in turn, first misinterprets, then is baffled by, her husband's state of mind. Thus a chain of actions based on a series of false models of the minds of others slides into disaster.

There have been a number of attempts to elucidate the brain bases of Theory of Mind. That there is a discrete neural network dedicated specifically to "social brain" processing seems a predominantly British hypothesis, promulgated notably by Christopher and Uta Frith of University College London and Helen Gallagher of Glasgow Caledonian University. The Friths speculate that ToM neural nets might have coalesced from a collection of more elementary skills by which creatures assign a kind of "aliveness" and intent to the actions of others in their packs, prides, herds, and so on. They propose that the dorsomedial PFC and the anterior paracingulate cortex, aka the rostral ACC (BA's 9 and 32), are central in this circuit that also includes some temporal and parietal lobe parts, high-level sensory areas, and the amygdala. But so far, only preliminary evidence suggests that these brain areas constitute a "neuroanatomical package," as Adolphs puts it, exclusively dedicated to Theory of Mind.

The Emory anthropologist James Rilling and colleagues have conducted a series of imaging studies to explore Theory of Mind. To stimulate thinking about the intent of others, they used two tasks: the Ultimatum and the Prisoner's Dilemma games. Popular in the testing devices in game theory and economics for decades, both arouse feelings about reciprocity, cooperation, fairness, and trustworthiness—as well as greed, selfishness, and cheating.

The investigators suspected the Ultimatum and Prisoner's Dilemma would elicit neural responses more powerful than those evoked by other experimental games such rock-paper-scissors where game players' decisions have less impact on personal self-interest. Participants were scanned as they played rounds of the games. In both games, brain regions that blazed in response to witnessing a "partner's" decision included the dorsomedial PFC, ACC (BA 9/32), and the right mid superior temporal sulcus (STS). The Ultimatum game, which involves evaluating the fairness of money exchanges, also revealed activity in brain areas not previously linked with ToM, such as the posterior cingulate cortex. The posterior cingulate, with its links to subcortical emotional systems, might regulate how we calibrate our emotions in response to feedback from other people.[12]

Although the Emory findings support the idea of a "social brain" network, with the medial PFC as a leading contender for a Theory of Mind center, these same networks are the ones we use to evaluate our own feelings (although judgments of others can selectively activate a network including the left lateral PFC).[13] Thus areas that light up when we assay our own emotions in a high-stakes poker game might also be engaged when we check the psychic bluffs of other players, the tells—body mannerisms and tics—that betray the strength of a competitor's hand.

Iowa's Steven Anderson admits to "never being real comfortable with the concept of a dedicated Theory of Mind circuitry. There are many complicated cognitive processes involved—perceptual abilities, flexibilities of thinking, sophisticated working-memory mechanisms. Theory of Mind can break down at many stages. But patients with orbital prefrontal damage do have an impairment in understanding the psychology of others. If we ask family members of these patients what's the biggest problem they face," he continues, "a lack of empathy is very high on that list. That's what's really devastating to spouses and children: 'He doesn't understand anymore when I am crying. He just doesn't get it.' When it comes to interpersonal relationships, boy, that's a big one! So I have no doubt there's something there. But this a hugely complicated psychological process, and it will probably be a huge circuit."

Jeremy Gray, too, finds the jury still out on Theory of Mind circuitry. "I'd be surprised if there were a completely dedicated system that does nothing but process social emotions. It wouldn't make sense, since we already have in place a system for processing emotions that could also be involved in the affective part of social interpretation. Since we have this hardware in place for doing these emotional jobs," he muses, "it would be like buying another machine to do something specialized when you have a general-purpose machine. In biology you usually draw on the resources you already have. It would be more relevant to understand how these brain regions work together, the functional network that forms along these pathways, in trying to guess how somebody else feels. Of course, if there is a dedicated network, and you discover it, you'll be famous forever! But almost everything in the brain turns out to be: 'Wow, it's more complex.'"

For the nitty-gritty cellular neuroanatomy of self and others, Gray suggests checking out John Allman at Caltech, who is investigating a special neuron. Called spindle cells, these fairly large neurons seem to work as travel agents for routing and integrating emotional and reasoning information, especially information about self and others. Allman and his colleagues claim to have found spindle cells in the frontoinsular cortex, mainly BA 24, a sector often regarded as part of the orbitofrontal PFC. He has observed these neurons in only two species—humans and African apes. And humans may have five to forty times more spindle cells than apes.

Area 24 appears to serve as an interface between emotion and cognition. Its lower sector helps regulate heart rate and blood pressure, as well as being involved in the production and recognition of facial expressions. Our experience of any given intense emotion, says Allman, whether love, fear, or happiness, is associated with lower area 24. The top of area 24, however, fires up when a person engages in a combination of emotional and reasoning tasks. The spindle cells in area 24 may be involved in registering the appropriateness of social events and transactions; they may be key components in a system that polls the self's feelings about a given experience and sends the results around the cortex. Spindle cells first appear around the fourth month of life, and gradually increase in number during the next two years—at the time a child's awareness of self and others expands.

Thinking about the thoughts of others can be like walking

through halls of mirrors. In a contest, for instance, a player's mental model of the opponent includes not merely a shared database of the game's goals and rules but also awareness that the opponent may know more than you know, knows you know this, and so on. The economist John Maynard Keynes pondered this form of iterated recursive social analysis ("I think that you think that I think . . ."), comparing a stock market investment to a newspaper competition for choosing the prettiest face from a group, the prize going to a picker whose choice matched the most popular choice. "It is not a case of choosing those which, to the best of one's judgment, are really the prettiest, nor even those which average opinion genuinely thinks the prettiest," he said. "We have reached the third degree where we devote our intelligences to anticipating what average opinion expects the average opinion to be. And there are some, I believe, who practice the fourth, fifth and higher degrees."[14]

The University of Michigan psychologist Trey Hedden wondered how deep goes this recursive power. Students playing two-person matrix games, he saw, possessed a default mental model about their partners that dynamically changed as new evidence about an opponent's moves and strategies came in. Player Bob tended to model the "mental state" of Danny only to the default stage, as if Danny would myopically not anticipate Bob's own action. This, then, is only a two-stage depth of anticipation of the other's moves. Heddon saw no stage three or four iterations.[15]

How far can one go in strategic interpersonal iterations, then? Even though game theory assumes that players pursue recursive reasoning indefinitely, and thus attribute the same strategic powers to one's competitors, it may be that humans are bound by the cognitive limits of our executive functions. Working memory is limited in capacity, as we've seen, to somewhere between four and seven chunks of information. Research into the processing of complex, recursively embedded sentences, furthermore, has shown that three or more levels of embedding tend to generate errors in comprehension and remembering.

The PFC and Violence: In Cold Blood?

Steven Anderson's cases, "A" and "B," could easily be members of *A Clockwork Orange*'s posse of rude Droogies. All could be exhibits of antisocial personality disorder (APD), the catchall psychiatric condition characterized by a pervasive disregard for social and moral rules, irresponsibility, lack of remorse, and impulsive violence. People with APD are those who blow up frogs as children, vandalize cemeteries, hurt people, and end up in the criminal justice system. Might individuals with defective prefrontal systems enter this world as "natural born" hell-raisers? Does early damage to the prefrontal network, moreover, "cause" general miscreant and criminal behavior? And, finally, could an abusive early environment transform a person born with a normal brain into someone indistinguishable from the brain-injured sociopath, into sociopathics such as "A" or "B"?

"The short answer to all that is yes," responds Steven Anderson. "Certainly people can be born with damage to these areas. We can image it clearly with MRI. What we're not studying are people with more subtle dysfunctioning in this system that we are not yet good at seeing. And if we were truly, truly evil people, we could take a perfectly healthy infant with an entirely normal brain and raise that child to be a sociopath. If raised in a psychopathogenic environment, someone can undoubtedly arrive at the same state."

Aggression and violence are built into living organisms. Take the fruit fly. At Harvard Medical School, the neurobiologist Edward Kravitz has shown the ubiquitous lab insect to be a ferocious fighting machine. Kravitz's team videotaped over a hundred boxing matches between young males of the species *Drosophila melanogaster*, each of which often lasted thirty minutes, with each fly lashing out at the other more than two thousand times (female fruit flies are effective pugilists, too). The Kravitz group is using these tiny, flying Mike Tysons to explore the genetic foundations of aggression.[16]

Or observe apes. Male chimpanzees sometimes gather and mill about in a loose pack until their bonding behavior heats up to a critical boil of aggression. At this point they form a battalionlike unit and set off on patrol, combing the borders of their territories for alien chimps to bludgeon to bloody pulps. For them, as for most species, a certain amount of aggression and potentially harmful risk-taking can yield big payoffs. But as with humans, some apes are uncommonly violent, and persistent over-the-top acts of aggression result in their ostracization, thus lack of access to females, breeding failure—even early death.[17]

It has been noted not infrequently that we humans are transfixed by violence. It is as if upon occasion we are culturally spellbound under a vast hypnosis of mayhem. We seek it out in our entertainment, and have it thrust upon us by our media, whether we want it or not, sometimes in almost orgiastic repetitions of horrifying images, such as fall of the World Trade Center towers. At the same time, we labor to demystify the "evil" of violence, to expunge the devil that "crawled into you," to find a rational explanation for the extremes of behavior. Brain scans reveal that when a normal human, even a professed pacifist, views film clips featuring ultraviolence and death, his or her visual cortex, limbic centers, and PFC start hopping with activity. As one fMRI scanee reported, "My brain was atwitter" when fed images of depravity, and he recalled E. M. Forster's observation that the "human mind is not a dignified organ."[18] This aggression meter in the brain that treats the representation of violence like neural candy craves more of it. So our media may be catering to a very real part of our unconscious neural selves.

Evidence increasingly suggests that violent, antisocial behavior has a neurobiological basis (although most people with brain disorders are not violent). Supreme Court debates on the death sentence for the mentally ill and for youths who commit heinous crime before age eighteen reflect growing consideration of brain development and mental illness as mitigating factors. Yet in 2003, Human Rights Watch reported that as many as one in six of the 2.1 million incarcerated Americans are seriously mentally ill. The study finds that as state hospitals close and prison populations have more than quadrupled in last three decades, jails and prisons have become the country's de facto mental health system.

Dorothy Otnow Lewis is a revolutionary thinker about criminal behavior. Her 1998 book *Guilty by Reason of Insanity* is a startling revelation of the neuropsychiatric status of death row convicts and other savage denizens of U.S. prisons. Lewis, a clinical professor at Yale's Child Study Center, and Jonathan Pincus, the chief of neurology at the VA Medical Center in Washington, D.C., and a neurology professor at Georgetown Medical School, have explored the brain-environment nexus of violent behavior. Lewis argues that most of the murderers executed on death row in the 1990s were so neurophysiologically impaired that they were not culpable for their crimes. Her recent data on the death row of Texas strengthen her previous findings.

Lewis has studied hundreds of murders, interviewed "celebrity" serial killers like Ted Bundy and Arthur Shawcross, and John Lennon's shooter, Mark David Chapman. She is quick to agree that most brain-damaged people are not violent, nor are most people with serious mental illness violent, nor are those abused as children. What does offer a prescription for violence, according to Lewis and Pincus, is a situation in which a child with brain abnormalities or injuries—especially in the prefrontal cortex—is raised in an acutely abusive environment. The abuse triggers impulses toward violence and aggression that an already weakened executive control system cannot compute or suppress.

Near the book's end, Lewis tells of her search for the one "pure" sociopath, an individual who is truly cold-blooded, remorseless, empathy-free, yet who is neurologically "normal." This sociopath will have no brain dysfunction nor a history of childhood abuse. At one point she thinks she has found this "ideal" specimen in the person of a prison executioner, an electric chair specialist. But upon interviewing this man at length, she discovers that this death-bringer, too, had suffered brain injury as a child and had grown up in a nightmarish household—a childhood whose memories he, for the most part, repressed. Lewis mulled over the moral question at the heart of his ghoulish profession: "Could anybody do it . . . ? I doubt it . . . Most of us are much too squeamish to kill another human being except in self-defense. It seems to take intense, repeated, intolerable pain early in life, and some sort of organic impairment or psychotic thinking to overcome that taboo."[19]

Jonathan Pincus's *Base Instincts: What Makes Killers Kill?* is a less compelling but physiologically well-grounded narrative of similar studies. Violent criminal behavior for Pincus, too, is the disastrous convergence of neurobiological and environmental factors. Of the 150 dangerous people he examined, 94 percent had suffered severe physical violence and/or sexual abuse as children and had brain dysfunction, usually in the frontal lobes. Many of these killers did not express remorse or comprehend the consequences of their deeds. They might acknowledge they did something wrong but didn't feel it as wrong. (This mind-set was on display in the 2005 Wichita "BTK" serial killer case. In pleading guilty to ten murders, Dennis Rader, the former Boy Scout leader and church president, spoke emotionlessly of his victims as "projects," explaining matter-of-factly of his first murder that "I had never strangled anyone before, so I really didn't know how much pressure you had to put on a person or how long it would take.")

Lewis's and Pincus's work discloses the murky state of evaluative procedures in the mental health and forensic worlds. About one young murderer, Pincus writes that a "perfunctory psychiatric assessment concluded he had 'antisocial personality disorder.'" In a raw surge of annoyance, Pincus adds, "That he had behaved antisocially was not in doubt. The issue was why. The useless tautology that equates antisocial acts with antisocial personality is the diagnostic redoubt of clinicians who do not wish to take the time to perform a thorough assessment or to think about the issue." Pincus suggests that few mental health experts seem to realize that symptoms associated with antisocial personality disorder can be signs of frontal lobe damage.[20]

Even though violence has been studied in psychiatry, social sciences, and psychology for over 150 years, rigorous investigations of the neural substrates of unwarranted aggression have been few and far between. Much spurious data, mainly from anecdotal stories of impulsive rage and cruelty, has led researchers into Dantean circles of blood and more blood. But by the end of the twentieth century, some investigators felt it was time to take stock of what they knew and what they didn't. In 1998, the Aspen Neurobehavioral Conference, an annual confab devoted to understanding issues related to mind and brain, convened a group to discuss the neural bases of violence. Bruce Price, the chief of neurology at Harvard Medical School, vividly remembers the meetings. "To see what they'd get, they put together neurologists, psychiatrists, neurophysiologists, Jesuit priests, lawyers, ethicists, a trauma surgeon, and a nurse. And they didn't let us out of this gorgeous room in Colorado for five days."

What they got was a written consensus, summarized in a 2001 paper by Price and others that concluded, indeed, violence can sometimes result from brain dysfunction, although social-environmental and genetic factors certainly contribute. They underscored the urgent need for intensive study of the brain-behavioral matrix of violence, particularly frontal lobe dysfunction, altered neurochemistry, neurometabolism, and the influence of heredity.[21] It was a start.

Price's original commitment to the brain bases of violent behavior began in the 1960s, when, as an idealistic, activist, naive college student, he taught a seminar for patients at what was then Bridgewater State Mental Hospital outside Boston. "I was very taken with Freud at the time, and thought he was immediately applicable to those on the sexually insane ward at Bridgewater," Price recalls with a touch of ruefulness. "I quickly learned Freud was irrelevant." During this time, a patient who attended Price's seminar was discharged after thirteen years on the ward. "I had the notion I was going to change the very nature of this gentleman," Price recounts. "When he was discharged, he had no place else to live so he came to stay in a dorm with my friends and me. I was caught up in the nurture-versus-nature idea. We didn't know much about it, but I'd just seen the great François Truffaut film *L'enfant sauvage*, a kind of Romulus and Remus story about this character who takes a child raised primitively and tried to civilize him. I had my own movie in the making.

"A couple of months later, this man pulled a gun and put it to the head of my best friend and said coldly, 'I could blow you away, and it would mean nothing to me.' The guy did not pull the trigger," Price hastens to add. "But there went my experiment. It turns out he had murdered before, and truly was this a remorseless sexual predator. And the last I heard, he had gone down to New York City, in all likelihood to become a contract killer. He is probably down deep in the East River by now." The young Bruce Price had clearly been in over his head, but the experience left him with an even fiercer desire to understand such people. "What made him so different? By the way, he was a bright guy. God, was he bright!"

In 1990, Price published his first report, in *Brain*, on two people who suffered bilateral PFC damage in childhood. Price, who examined them when they were twenty-five and thirty, saw basically the same constellation of behavioral problems as Anderson's team observed in "A" and "B." "The two were," says Price, "remarkably compulsive, didn't learn from negative experiences, and didn't have the ability to take another person's perspective. Their 'sympathy center,' if you will, was nonexistent. They committed one social foul-up after the next."

Price went on to investigate and treat other forms of mental illness. But after the Aspen Conference, it was clear that knowledge about the frontal lobes' role in violent behavior was chaotic at best. Price set out to "collect his thoughts" by writing up the state of the data. In this meta-analysis, an attempt to bring coherence and perspective through a review of all preexisting studies, he collaborated with Montgomery Brower, then a chief resident in neuropsychiatry under his direction in 1998 at Harvard's McLean Hospital. Brower has served as deputy medical director of Massachusetts's maximum-security forensic psychiatric hospital and as senior psychiatrist on a Massachusetts forensic mental health inpatient service. As a forensic psychiatrist, he deploys not only neuroscientific training but also his experience at the crossroads of law and mental health, diagnosing and treating some of the sickest and most violent people in the state of Massachusetts. Now in private practice, Brower gives expert witness testimony in the courts and treats juveniles in secure detention.

The Meta-Analysis of Mayhem

Brower and Price scoured the brain-violence literature back to 1835, to the time of the first attempts to categorize and measure antisocial personality traits following frontal lobe injury. Later, large studies of war veterans with head wounds emphasized the association between frontal lobe lesions and aggressive or antisocial behavior, although the amount of actual crime appeared small. The Vietnam Head Injury Study, for example, found that veterans with lesions limited to the frontal lobe, especially mediofrontal and orbitofrontal injury, showed more aggressive and violent behaviors than did vets with non-frontal-lobe injury. These historical studies were done after the fact of injury, however, and did not control well for known violence risks factors such as prior criminal record, history of drug abuse, economic status, or previous psychiatric symptoms. (The Iraq wounded will, regrettably, provide more data.)

"It was a massive amount of work," Brower admits about the paper, "Neuropsychiatry of Frontal Lobe Dysfunction in Violent and Criminal Behavior: A Critical Review."[22] "There had been no definitions, no sense of what the studies were really looking at. They were all over the place." Brower and Price defined "aggression" as any threatening or physically assaultive behavior directed at persons or the environment, and "violence" as actions that inflict physical harm in violation of social norms. They found that research by different disciplines used very different tools and theoretical perspectives. So they separated the disciplines into categories. One was "garden variety researchers," medical doctors and psychologists describing case series or reporting individual cases, evaluating symptoms, attempting to diagnose and treat disorders, and putting things into clinically meaningful categories.

Another category, the neuropsychologists, approached the problem of violence with testing tools that to varying degrees were valid and reliable measures of cognitive capacities or abilities. Some tests revealed the links between violent criminal behavior and frontal lobe dysfunction to be weak at best. Others concluded that attention deficit hyperactivity disorder (ADHD) probably accounted for many problems in executive function that were tied to adolescent aggression and troublemaking. One study focused on the virtually self-evident idea that ADHD might exacerbate aggression in these teens. Another study found that low executive function test scores predicted not only which boys ages 10 to 12 admitted to aggression, but also which boys' fathers had histories of substance abuse and social problems. Criticized for not taking ADHD into consideration, the investigators conducted a follow-up study of aggression in troublemaking teenage girls that excluded girls with ADHD. Low executive function scores still correlated with physically aggressive antisocial behavior.

Neuropsychologists tried to tease apart the relationship between specific thinking problems and aggressive behavior—to determine whether violent behavior can be tied to some cognitive deficits but not others. Swedish scientists led by Asa Bergvall of Göteborg University tested a group of criminally violent men undergoing court-ordered psychiatric examinations. Their tests included measurements of planning ability, working memory, and attentional set-shifting. A set-shift task required them to focus attention on one visual dimension, shapes, then reverse attention to focus on lines, a shift from one perceptual dimension to another. The men's performances were compared both to participants with marginal mental retardation and to normal people. Violent offenders performed as well as did the normal subjects on tasks of spatial and figurative working memory, as well as on a test for planning.[23]

They were markedly impaired, though, in attentional set-shifting, committing three times as many errors on average as either the normal or mentally retarded subjects. Violent offenders performed well on every task that taps prefrontal function but this one, commented Bergvall. "In that, it was as if they were retarded." At the heart of the deficit was their stunted ability to change their focus in response to fluctuations in events or experiences around them. They had trouble altering their perspective of the world when that world changed before them.

Many in this violent offender group had childhood histories of hyperactivity, poor attention, and impulsivity—conditions that could be tagged as ADHD, or even less instructively, as "developmental coordination disorder" or "conduct disorder." (One can imagine a couple of gangbusters saying, "Dude, your problem is you've got 'conduct disorder'—which means you've got problems when you try to conduct yourself. Don't you dig that a tire iron is not a socially acceptable instrument for expressing your feelings?") Because these conditions often persist into adulthood, they are risk factors for adult criminality. "It is quite possible that the cognitive deficits we note in the present study constitute adult forms of these developmental psychopathologies," adds Bergvall.

This research has data-based, statistical power, Brower explains. "That's the plus side with the neuropsychological stuff. But on the minus side, what does it bring to the physician: the 'Okay, tests show that Mr. X is impaired on a set of cognitive performance standards.' Does it translate into something that I as a doctor walking into the room and seeing him will recognize as a syndrome I can treat? It's hard for a physician to apply these abstractions culled from lab experiments to the flesh-and-blood individuals they examine. And even if the psychologist has a patient whom he's identified as having a problem with his executive function," Brower argues, "will that rise to a significant or treatable syndrome? I may walk into a room and think, 'Gee, Mr. X has some problems focusing his attention; he's a little impulsive.' But do I necessarily recognize this as a full-blown antisocial personality disorder? Or do I say, 'This person has some cognitive impairment, some impulse control problems, but he's not really violent or mentally ill'?"

The third group whose work Brower and Price scrutinized was the newest: the brain imagers. Brain mapping of people with personality problems basically began around 2000, about the time Richard Davidson promulgated his "faulty circuit wiring" theory. The Wisconsin group analyzed imaging data from more than five hundred violent individuals, including murderers, childhood brain trauma victims, and those diagnosed with antisocial personality disorder. The evidence indicated such people might have defects in circuitry involving not only the PFC, but the ACC and the amygdala, among other subcortical structures. Wiring glitches between such brain areas might weaken one's ability to regulate or control negative emotions and "read" the cues in other individuals that most of us use to rein in our aggression.

Davidson and colleagues focused specifically on impulsivity—spontaneous, unplanned aggression where one "acts without thinking." Positing that a predisposition to impulsive acts of violence is a consequence of impaired emotional regulation within prefrontal and limbic circuitry, Davidson embedded this concept of aggression within his "affective style" hypothesis. Each of us utilizes a unique set of parameters to govern our emotional reactivity. A propensity for impulsive aggression is part of the negativity set—such as anger, distress, and agitation—that implicitly seems to "dismiss" the future consequences of behaving aggressively as a matter of concern.

Broadly defined, impulsivity refers to behavior that takes place in the absence of conscious decision-making. Bruce Price asks, "Among most of us rather civil people, who hasn't had the urge to just throttle somebody? But most of us have the good sense to analyze that urge, judge it, and say, 'Boy, would that get me in trouble!' When we were young and temperamental, our parents said, 'Look, count to ten . . .' That was good advice, because it gives the prefrontal cortex time to do the analysis and suppress the heavy impulse. In these antisocial personality types, there could well be misregulation between the limbic and prefrontal areas." Faulty wiring might also result in much slower "extinction of nasty responses as well; say, the perpetuation of spontaneous rage as smoldering resentment. The impulsively aggressive person may be quicker to feel negative emotions, and slower to get over them." Or be unable to reappraise a situation.

Neuroimaging data are pointing to selective PFC dysfunction as a root of impulsive aggression. In the behaviors of people with orbitofrontal/ventromedial—as opposed to dorsolateral—PFC damage, problems of impulse control and disinhibition are usually clear-cut, to the point where, says Brower, "if you were to sit down with a person with a significant OFC lesion and listen to his conversation, you would have little trouble realizing he was inappropriate in his social interaction, was highly distractible, jumping from one topic to another, lacking insight, unable to manage his personal boundaries, and maybe impulsively obscene. He would display a quality of superficiality and lack of depth and continuity in his social interaction. But for a person with dorsolateral dysfunction, you might need a battery of neuropsychological tests to really be sure something was not normal."

Raine's World

The few imaging studies of the aggressive and violent have consistently found frontal abnormalities. But with lack of replication one wonders what to make of it. The most notable imaging work on the prefrontal bases of violence has come from Adrian Raine's lab. A curly-haired Brit with a certain sangfroid, Raine has hands-on credentials in the criminal justice system. After receiving his PhD in the United Kingdom, he served for four years as a psychologist in top security prisons in England. Raine immigrated to Southern California in 1987, he once stated, because "in addition to the good weather, there were plenty of murderers." In 1993, he published a comprehensive text, *The Psychopathology of Crime: Criminal Behavior as a Clinical Disorder*, detailing what was then known about the brain bases of antisocial behavior. By the end of the 1990s, at the University of Southern California, where he is the Robert G. Wright Professor in the Department of Psychology, he had teamed up with the UC Irvine neuroimaging pioneer Monte Buchsbaum to peer into the brains of killers.

The series of studies Raine spearheaded is unique in the annals of neuroscience: they were the first (and perhaps to date, only) experiments to explore the brain bases of behavior in large groups of dangerously aggressive criminal offenders. In 1997, Raine's team studied thirty-nine men and two women who had been indicted in California for murder or manslaughter. In police custody, all had been referred to the UC Irvine imaging center for various reasons relating to their defense strategies. All had either pled not guilty, not guilty by reason of insanity (NGRI), or incompetent to stand trial. Justification for the imaging referrals ran the gamut: schizophrenia, history of head injury or other brain damage, substance abuse, epilepsy, affective disorder, hyperactivity, learning disorder, paranoia, and others. Raine and colleagues PET-scanned this wild bunch after they had been medication-free for two weeks, along with a matched control group of noncriminal subjects.[24]

The scans revealed that the murderers had significantly lower rates of glucose metabolism than did controls in both the lateral and middle PFC in both hemispheres, as well as in the parietal, corpus callosum, amygdala, and other subcortical regions. These findings, Raine claimed, provide support for the dysfunctional PFC theory of violence: that prefrontal damage could result in "acquired sociopathy," that the PFC is no longer able to keep a lid on the maladaptive urges welling up from the "cellar." Glaringly obvious was the fact that these "NGRIs" composed only a subgroup of violent offenders, and the sluggish metabolism in their PFCs could not be generalized to other types of violent criminals, nonviolent criminals, or noncriminal but highly aggressive people. Commenting on the Raine findings, Dorothy Otnow Lewis remarked that she would very much doubt that any of these impairments would create a murderer. But if one of them had also been horribly abused, it could be devastating.[25]

Next, Raine separated the same bunch into two groups of aggressors: predatory psychopaths, and "affective" impulsive individuals with antisocial personality disorder. They were either controlled, purposeful, cold-blooded killers, or hair-trigger and rage-propelled, foaming-at-the-mouth marauders. Using a forensic rating system to categorize the two types, he took into account assessments from mental health professionals, criminal transcripts, conversations with prosecutors and defense attorneys, medical records, newspaper stories, and more. It was impossible to pigeonhole every offender into either pure category, so he retained only those murderers who rated "strongly predatory" or "strongly affective," leaving fifteen in the predatory and nine in the affective-impulsive camp.

Here Raine was in fact trying to clarify persistent, murky distinctions between APD and psychopathy. The *Diagnostic and Statistical Manual of Mental Disorders*, Fourth Edition (DSM-IV), the American Psychiatric Association's bible, offers diagnoses of conduct disorder for children and ADP for adults. But their parameters are so vague that they tend to corral diverse groups of people and are fairly worthless. There is no easy way to ascertain who is truly suffering from APD, much less who is a predatory psychopath. Differences between predatory and impulsive antisocial personality disordered individuals have been most carefully studied by the psychologist and criminologist Robert Hare of the University of British Columbia. Hare has presented an alternative, perhaps more precise method, called the revised Psychopathy Checklist, that rates twenty different behavioral, emotional, and interpersonal characteristics and tabulates their "score" through extensive reviews and interviews.

Unlike impulsive antisocial offenders, psychopaths often use charm, intimidation, and, if necessary, violence to attain their goals, Hare said. Since they are three to four times as likely to offend again, it is even more important to distinguish them from APD people in forensic populations. Psychopathy leaves a person remorseless, guiltless, without empathy or inhibition against using deceit to facilitate harming others. For these shallow but superficially charming individuals, he said, "social rules have no constraining force, and the idea of a common good is merely a puzzling and inconvenient abstraction."[26] "They have a self-centeredness—call it narcissism if you will. They are angry, paranoid individuals, with a diminished sense of guilt; that is, their victims deserve it," adds Bruce Price.

Psychopaths make up only about 1 percent of a population, but they account for a large proportion of serious crime, violence, and social distress in every culture. And while they account for 15 to 20 percent of prison populations, persons with psychopathic tendencies, Hare believes, are able to blend in well in today's "camouflage society" where "greed is good," success by any means including deceit, and style over substance are accepted if not openly valued. For Hare, psychopaths are everywhere—in the form of drug dealers, swindlers, con men, high-pressure salesmen and stock pushers, mercenaries, corrupt politicians, unethical lawyers and doctors, terrorists, and cult leaders. Inspired by Hare, others have gone so far as to talk about a "culture of corporate psychopathy," featuring CEOs and executives who meet the criteria for diagnosis. Mr. Skilling, please enter the scanner.

Matthew Stanford, of Baylor University's Department of Psychology and Neuroscience, also wanted to clarify the differences between impulsive and premeditative aggressors. The latter, he found, tend to control and channel their aggression to achieve their goals. Impulsive aggressors, on the other hand, may describe themselves as "Jekyll and Hydes," feeling themselves to be "two different people." While both groups are "very angry and hostile," impulsive APD people performed poorly on neuropsychological tests and had trouble planning and strategizing. Premeditative aggressors, however, typically perform normally on these cognitive tasks.

Adrian Raine wondered if PFC metabolic activity differed in the two groups. His PET data did, in fact, show that impulsive murderers had reduced prefrontal metabolic activity when compared to normal volunteers. The impulsive killers' PFC dysfunction might well leave them at the mercy of their overstimulated subcortical drives. Prefrontal metabolism in "predators," however, resembled brain activity of the law-abiding controls. One predator in his group, who Raine referred to in a lecture at Dartmouth in 2001, had killed an astounding sixty-four people in twelve years. "He had excellent prefrontal functioning, high glucose metabolism. He had to have something going that allowed him to escape detection. He was not an impulsive killer."

That the cool predators appeared to have normal PFC function squares with Hare's and others' notion that these offenders have the cognitive reasoning power to elaborately strategize, anticipate consequences, and control their aggressive behavior to achieve their horrific goals. John Wayne Gacy, the "killer clown," the legendary Ted Bundy, and even the BTK killer could qualify as intelligent planners—not to mention the genius of, albeit fictional, Hannibal Lecter. But this leaves the gaping question of why such people become walking nightmares.

What the premeditative psychopaths and impulsive slayers in Raine's study shared was excessive metabolic activity in the midbrain and thalamus and in the limbic areas of the amygdala and hippocampus in the right hemisphere. The overstoked limbic fires in both sets of murderers were consistent with other human and animal findings in which limbic hyperactivity correlates with aggressive behavior. Raine suggests that part of the explanation of the predators' proclivities lies in this abnormal right subcortical activity. While they have enough prefrontal juice to regulate their aggression in deceiving and manipulating others, something excessive from the dank subterranean caves of their psyches boils up to steer PFC executive forces toward gruesome premeditated violence.

In the psychopath there may be a kind of violence circuit. Cues in the adult's environment might trigger suppressed, trauma-laden memories of a brutal childhood, which in turn ignite powerful subcortical activity. This excessive activity, stoked by reexperienced and intensified negative emotions, could drive the psychopathic PFC to formulate a "plan of action." Of course this hypothesis leads to the next question: just how does this putative subcortically driven compulsion to acts of violence co-opt higher-order PFC operations? Jonathan Pincus finds obsessive-compulsive "rule-system" characteristics in some methodological killers he has observed. This compulsive quality, he says, may prevent serial killers from being violent when they cannot "do it the right way," such as when they are in prison. "The need for the right time, place and victim stamps these acts as bizarre, and protects other inmates from the serial killer's vulnerabilities to violent action."[27]

At the Dartmouth lecture, Raine put another PET-scan image up on the big screen. "See, here's a second individual with good frontal function, like the man who killed sixty-four people," he said gesturing toward the slide. There is, he points out, the same pattern of high activity in the temporal lobe as well. Raine skips a beat, then added, "This is my brain scan. I've never been convicted of a homicide, and I don't want to be locked up!" Raine performed this gimmick to stress that single measures—whether biological or social—can never alone diagnose or predict a killer. His own findings, he admitted to his audience, are crude answers to complex questions. And despite their occasional brilliance, serial killers do lose control and make mistakes. Hannibal Lecter not withstanding, most violent psychopaths can keep it together for only so long before they start to unravel—as if their executive functions can finally no longer harness the chaos welling up from suppressed memories and emotions.

Sociologists have long held that poverty and poor education are fertile incubators of crime. But such notions don't explain the existence of unrepentant silver-spoon criminals—from white-collar CEOs to Timothy McVeigh, Ted Bundy, the Unabomber Ted Kaczynski, and perhaps to middle-class, well-educated terrorists throughout the world. If a violent offender comes from a verifiably good home, would the wellspring of his or her violence have to arise more from neurobiology than nurture? After tracing the early histories of the same sets of murderers, Raine reexamined the scanning data from the perspective of family background: benign and comfortable versus traumatic and deprived.

Raine's band of investigators combed court records, attorney interviews, and psychological and medical records for clues illuminating their subjects' upbringing. Did they suffer physical neglect or abuse? Did they have criminal parents or grow up in a household wracked by conflict or extreme poverty? The scientists rated the severity of risks. Repeated physical beatings or sexual abuse in childhood got a high score on the subject's psychosocial deprivation chart. Slightly fewer points were allotted to evidence of intense family conflict and placement in a foster home, and so on, with deprivations such as poverty getting lesser points. Of the forty-one murderers, only twelve had suffered severe psychosocial abuse and deprivation. The remaining offenders had experienced minimal abuse or none. Neither group had higher incidence of head injury than the other.[28]

The PET-scan data clearly showed that only those killers rated as coming from fairly affluent social strata, and with competent, benign parents, had the telltale lower than normal prefrontal metabolic rates. Those of murderers from homes riven by conflict, deprivation, and abuse approached normal. Compared to the severely deprived subjects, murderers from good backgrounds averaged 4.7 percent less metabolic activity in the lateral and medial PFC. More specifically, the orbitofrontal cortex on the right side of the "good family" murderers showed even poorer functioning—14 percent less activity compared both to normals and to "bad family" murderers.

Raine hypothesized that offenders from very deprived backgrounds tend to commit violence for more psychosocial reasons such as childhood abuse, parental criminality, or family strife. But with killers from good backgrounds where the social push toward violence would be minimal, brain abnormalities become a more powerful explanation for their offenses. The good parents of unmanageably aggressive, problem children may wonder what they did wrong, said Raine, but it may be absolutely nothing. A neural defect may be the culprit. This finding may begin to clarify why some murderers seem to be products of their early environment, while others defy their upbringing. Consider, for instance, the differences between serial killers such as Robert Alton Harris and Jeffrey Dahmer. Harris, says Raine, "was battered from pillar to post all his life," raised by a father of similar brutality and an alcoholic mother. Gene mutation, not trauma-based dysfunction, may have played a relatively greater role with Dahmer—the notorious killer of seventeen young men and eater of body parts—who by most accounts came from a relatively normal home.

One reason we don't all go around carving up one another with steak knives or worse is, Raine thinks, because most of us learn fear conditioning such as we get when we're smacked or grounded by parents for petty theft or playground assault. We learn to associate antisocial behavior and punishment, and we experience the negative reinforcement of physically unpleasant symptoms when we contemplate a wrongful act. But someone with a frontal lobe deficit, especially in the OFC, as we've seen, may find it hard to make the associations, or feel the unpleasant bodily affects. Thus the OFC, argues Raine, is the prefrontal zone of relevance in violence. Underactivity in the right OFC in murderers from good backgrounds might underlie their reduced responsiveness to aversive emotional conditions.

Raine's data imply that in the set of murderers from good homes at least, the individuals were probably born with the brain deficits. But as with other phases of this study of forty-one NGRIs, the results cannot be generalized to other violent groups. If diminished glucose metabolism in PFC sectors predisposes a person to murder, then tackling the causes of these brain dysfunctions and treating them might help reduce violence. But, Raine hastens to reassert, few frontally impaired people go on to commit crimes— PFC dysfunction is neither a necessary nor a sufficient condition for violence. In demonstrating that murderers who come from relatively normal social milieus may be more likely than murderers from less socially privileged conditions to have PFC deficits, Raine's findings challenge assumptions made in forensic settings that violent offenders from normal backgrounds are neurobiologically "normal," whereas those exposed to abnormal early conditions are more likely to be neurobiologically deficient. Indeed, his data turn these assumptions upside down.

A number of forensic brain scientists pleaded to keep Timothy McVeigh alive, arguing that he, like few other world-class killers at the time, represented the closest thing to a window into the brain of violence, and that his death was tantamount to destroying evidence.

"Well, I wouldn't go that far!" Bruce Price hastens to say, adding that McVeigh himself rejected any scientific probes of his psyche or brain, in life or death, with the acquiescence of the U.S. government. "Of course there are legal and ethical problems," Price continues, "but why shouldn't we take pictures of their brains? Give them psychological and physiological tests? What switch do they have, or don't have, that is different from the rest of us? Why can't we try to see what makes them tick and try to apply that knowledge to help others? Some of these people are just built for useful research! Take McVeigh. We should have had a crack at his brain. It might have been normal—but that would be useful to know as well."

It speaks to a powerful cognitive dissonance that many Americans are so riveted by killers, yet harbor such a strong prejudice against keeping them alive long enough to understand anything about them. "Now the public as well as people in the professions, even neurobiology, psychiatry more so, are recognizing that not only things like language, memory, vision and motor movements are attributes of the brain, but also reasoning, executive function, empathy, sympathy, insight, awareness—these are all attributes of the brain. Since that's true, let's study brain in all its manifestations. That's kind of a no-brainer," says Price.

Adrian Raine also wanted to explore PFC function in people with antisocial personality disorder but who did not also suffer from drug addiction/substance abuse or mental illness. It was a distinction that had not been made; such types had all been lumped together. Raine also wanted to muster subjects from somewhere other than prisons and forensic psychiatric wards, where drug-taking, brain trauma, and psychopathy are often commingled in a single individual. To recruit volunteers, Raine's team fanned out into the Los Angeles underclass community—specifically to five temporary employment agencies. The temp agencies were known magnets for losers—people with relatively high rates of violent behavior, who tended to disregard the law, lie, pick fights, and get fired from one job after another. "Anyone who wanted to volunteer could," Raine says. From the batch of volunteers rounded up, the researchers put together three groups: twenty-one men with a relatively "pure" diagnosis of antisocial personality disorder; thirty-four men who were neither APD cases nor alcohol or drug abusers, who would serve as the controls; and twenty-seven men with substance dependence—either drug or alcohol addiction—but who did not have APD. All were between twenty-one and forty-five years old.

Included within the parameters of violence were acts that caused bodily injury or trauma, or were life-threatening, such as a history of attacking a wife, girlfriend, or stranger causing bruising or bleeding; rape; using a weapon in a fight; using force or a weapon to rob; firing a gun at someone; attempted murder; or murder itself. The APD group had committed significantly more serious violent crimes than the control or drug-addicted groups. All the information the scientists gathered from the men was confidential, and, Raine added, it seemed that this was the first time many of them felt free to talk about what they'd done.

Like Damasio's acquired sociopathy patients, the APD cohort from the streets of L.A. also showed abnormal bodily reactions, such as lower sweat (skin conductance) response and reduced heart rates during a stressful experience. The stress test involved asking each man to list his faults and describe them in front of a video camera. If lower than normal stress response is a sign of general lower levels of arousal, then a person's antisocial acts might be an attempt to compensate for lack of autonomic stimulation, Raine hypothesized. After each man was scanned with structural MRI to determine the tissue volume in his PFC, Raine saw that, compared to the other groups, the APDs had reductions in PFC gray matter—the neurons—but not in white matter, the myelinated nerve fibers linking neuronal groups. The reduction was 11 to 14 percent, or 10 cubic centimeters, about the equivalent of two teaspoons of gray matter, Raine said: "subtle but very real, striking differences."

Published in 2000 in *Archives of General Psychiatry*, these findings provided the first hard evidence that antisocial ne'er-do-wells may have a structural brain deficit. "Neurological research has shown that when there is structural damage to the prefrontal cortex, antisocial, poorly regulated, sociopathic-like behavior can result," Raine wrote. "Our findings show that these same brain deficits are found in antisocial individuals in the community, and they are independent of other risk factors for crime."[29] Raine again supplied the caveat that PFC impairments alone cannot explain APD and aggression, and indeed dysfunction involving more extensive neural networks—interacting with psychological and social problems—is likely to be the cause of aberrant aggressive behavior. But he, as others, maintained that the more we understand the neural substrates of violence, and can integrate that knowledge with socioeconomic and family causes, the better we will be able to predict for violent behavior and know where to target our intervention resources. We need a way to know which children will grow up to do the greatest harm to society.[30]

Raine offers a number of therapeutic strategies for children—the 5 percent who will commit 50 percent of the crime and violence later in life. Such strategies include cognitive behavioral training, similar to techniques known to help phobics, and biofeedback, training children to control their own arousal levels. Children might be encouraged to channel their energies into safe activities that would satisfy their stimulation-seeking and aggressive proclivities, steering them away from the antisocial, criminal life into which they are in danger of tumbling. Society needs fearless people, he added, noting that bomb disposal experts have low resting heart rates. People with low arousal might contribute to society as firefighters, test pilots, and search-and-rescue workers—if they are identified and helped in time. Society, in turn, might herald these people as heroes, positively reinforcing their struggle to maintain control and, perhaps most important, including rather than excluding them.

Indeed, he admits that these findings do not show a cause-and-effect relationship: some APD people might have significantly more PFC gray matter volume than people who don't have APD, and vice versa. Certainly, taken out of context, identifying reduced PFC volume could become a misdiagnosis of APD.

Monty Brower speaks as a physician wrestling with diagnosing and treating devastatingly troubled individuals. Raine, Brower notes, scans numbers of antisocial people and, in his well-controlled study, finds that an 11 percent reduction in prefrontal gray matter is associated with this violent behavior. He finds good correlation with multiple measures. "Great!" Bower exclaims. "But if I were to send any person I thought had antisocial personality disorder to get a scan, that MRI can neither confirm or refute his having APD. It's very unlikely a single scan is going to pick up this eleven percent difference that Raine has found statistically over a large group of people.

"I wonder if what we're discovering is something we already knew," Brower muses, "that people are, to some extent, the products of their environment, development experiences, varying opportunities, and accidents. We also understand we come into this world with an individual biology that is a unique inheritance from the human race and from our own particular parents. So we all have emotional vulnerabilities with which we enter into the world, and experiences that are variously health- or disease-promoting. And we all have a commonsense idea that somewhere there's a threshold where one person's behavior is so rooted in pathology that we justly and fairly cannot hold him to the same standards of responsibility we might someone else. But it's far from proven or justified, either on scientific or philosophical grounds, that an eleven percent difference in the size of the prefrontal cortex represents that threshold."

Raine's imaging data, however, underscores the big question: Does "John Smith's" personality disorder have something to do with his biology? Is his subtle neural disorder significant because it leads to his dysfunctional behavior? If so, asks Brower, "should the physician try to figure out where this is, so that he or she can treat it? Raine's argument is: criminal behavior, at least from certain perspectives, meets the criteria for being a brain disorder. The implication is that if we find data that suggest a biological problem, then we're not talking about something that can be understood in traditional legal and moral terms, as something that is subject to censure and punishment. It is then a disorder that is in need of treatment."

Still the question persists: is this really a mental illness that we need to understand like, say, schizophrenia, where clearly we have treatments and interventions? "If you ask anyone on the street if there's something wrong with a person suffering from schizophrenia," adds Brower, "he will have no problem telling you there is. But if you ask that same person about the individual with antisocial personality disorder, then your man is likely to say the guy's really an asshole."

This is where the forensic perspective is lacking in the neuroimaging studies, which are rarely sufficiently analyzed and clarified. "Adrian Raine has been something of an exception in this regard," adds Brower. "Over time, as his studies have been criticized on these grounds, he has made good-faith attempts to make his definitions of aggression and violence more clearly data-based, and to more clearly define the nature of the behaviors he's talking about. So he deserves credit for showing he's cognizant of these issues and is trying to make his data reflect some of those concerns." And that's the state of affairs right now.

The Brain-Drug-Violence Connection

The neural signposts leading to impulsive violent behavior point to one major neurotransmitter: serotonin. Levels of aggression, posits the current "serotonin hypothesis," are likely to be affected by this ubiquitous brain chemical and the genes regulating its metabolism. Serotonin, or 5-HT, has reached celebrity status, given the popularity of Prozac and the family of selective serotonin reuptake inhibitors (SSRIs) that have revolutionized the treatment of depression and other mood disorders. Nonetheless, SSRI modes of action on their targets—the serotonin transporter protein—and how 5-HT works remains something of a mystery. A mountain of evidence now links low serotonin levels to impulsive and disruptive behavior.

"Does that mean cause?" Bruce Price asks rhetorically. "No. But there is an association, a low serotonin syndrome, if you will." The association holds true among many species. Lab knockout mice, bred without genes critical to serotonin function, can be remarkably anxious, vicious, and mean. In primates, reduced serotonin levels lead to intense, unrestrained aggression. Researchers link low serotonin levels in humans with violent suicide attempts, impulsive aggression, and unpremeditated arson. Yet not all impulsively aggressive people have lower than usual serotonin levels, and certainly not all people with low serotonin are impulsively violent.[31]

The serotonin system is one of several remarkably complex sets of instruments in the brain's neurochemical symphony, with many receptor types identified, and many more not. Serotonin function in the PFC is far from being understood, although it appears to play a bigger role in orbitofrontal mood and evaluative operations than in the executive dorsolateral PFC. Performance on working-memory and planning skills, for instance, shows little effect if serotonin is diminished. But performance on gambling tasks that use orbitofrontal circuits can be impaired when players have depleted serotonin levels. The performances of low-5-HT players in some cases mimicked the gambling follies of people with OFC damage.[32] Impulsive behavior is also highly correlated with depression, obsessive-compulsive and other personality disorders, and drug addiction—all of which overlap in serotonin system–linked clusters—making it unlikely that serotonin function is specific to only one disorder.

Serotonin's primary effect in the brain is inhibitory. It puts the lid on excitatory neurochemicals by boosting the action of GABA (gamma-aminobutyric acid), the ubiquitous inhibitory brain chemical. Serotonin may work in reciprocal fashion with norepinephrine as well. While norepinephrine facilitates spur-of-the-moment action, serotonin suppresses it. The serotonin system also has an excitatory role, stimulating the release of dopamine, essential for PFC function in its reward-driving powers. Given that the PFC, especially the OFC/ventromedial areas, is an overseer of impulsive actions, the serotonin system is probably one of the brain CEO's major chemical enforcers, exerting an inhibitory control over impulsive emotions. Yet where's the evidence?

Much serotonin research in people involves the neurotransmitter's chemical precursors in diet. A recent British study found that improving serotonin precursors in the diet of inmates in a maximum-security prison reduced their bad behavior. The improved diet in the Aylesbury prison included supplements of omega-3 fatty acids that raise brain levels of serotonin. Half of a group of eighteen- to twenty-one-year-old prisoners received the supplements, and half got placebos. Those getting the supplements committed 25 percent fewer jailhouse offenses than those who did not. And within two weeks of the treatment, the supplement takers committed 35 percent fewer offences than they did before starting the diet, compared with 7 percent on the placebos.

A decade ago Gerard Moeller, a psychiatrist at the University of Texas Medical School at Houston, was searching for clues to the relationship between brain levels of serotonin and aggression. In a classic study, he asked male volunteers to live for ten days on a milkshake regimen that lowered their blood levels of tryptophan, a precursor molecule of serotonin. The synthesis of serotonin depends directly on the availability of tryptophan, and by limiting it, brain serotonin supplies will drain away. After living on these low-tryptophan shakes, the men scored higher than previously on lab tests of aggression. (Since then, much has been made in the popular press about the sleep-inducing properties of turkey, which is rich in tryptophan. Milk, yogurt, eggs, meat, nuts, beans, fish, and cheese—cheddar, Gruyère, and Swiss—are also rich in tryptophan. And these foods have been touted as natural antidepressants. (Thanksgiving has become our premier tryptophan holiday.)[33]

More recently, Moeller's group tried tryptophan manipulation on twelve women volunteers to measure laboratory-induced aggression in the XX sex. The women's irritation and aggression quotients changed and were defined by the tryptophan diet, and their emotional fuses grew shortest on the tryptophan-depleted diet, just as in the men's study.[34]

And how might testosterone affect aggression—after all, hasn't being male been the number one excuse for behaving aggressively? Many studies on testosterone activity show a relation between high plasma levels of the male sex hormone and a tendency toward aggression. Some investigators think the interaction between low serotonin and high testosterone levels is a brew for inciting aggressive behavior. Testosterone may modulate serotonin receptors in ways directly affecting aggression, fear, and anxiety. Anecdotal evidence concerning the mood swings of bodybuilders on anabolic steroids—the tendency toward aggression and paranoia—suggests this may be the case.

As always, the story of a neurotransmitter family leads to upstream receptors and the genes that regulate them. Back in the 1990s David Goldman, at the National Institute on Alcohol Abuse and Alcoholism (NIAAA) was scrutinizing genetic variants in serotonin receptor 5-HT7. A genetic anomaly was discovered in 81 Finnish alcoholics with APD and in only 1 of 232 "well-balanced" Finnish control subjects. Goldman admitted at the time that many researchers were loath to talk publicly about the genetic underpinnings of violence, fearing to walk this political minefield. And things have not changed much. But Goldman, a bit of a cowboy in a lab coat, basically welcomed the chance to get his findings out in the open. He is currently investigating whether the gene variant for 5-HT7 is found in other populations, and if so, whether it can be linked to alcoholism and APD. If an amino acid substitution in the 5-HT7 receptor leads to its altered function, he says, "it may be a useful clue for developing new therapies and diagnostic techniques."

Explorers of the genetic bases of aggression inevitably turn to the serotonin system in primate evolution. Investigators recently traced back 25 million years one variant, or allele, that is known to predispose humans to impulsive, aggressive behavior, a so-called warrior gene.[35] The gene, found on the X chromosome in apes and monkeys, encodes the enzyme monoamine oxidase A (MAO A), which breaks down multiple neurotransmitters—including norepinephrine, dopamine, and serotonin—and so prevents excessive amounts of them from lingering in the synapse.

The MAO A gene is not to be confused with the 5HTT transporter. These are two different genes. The 5HTT transporter gene's function involves serotonin and acts on transportation of the serotonin neurotransmitter substance between synapses. In contrast, MAO A is a gene involved in the function of monoamine oxidase, a substance that oxidases (clears away) excess neurotransmitters of all types from the gap between synapses. MAO A is not specifically related to the neurotransmitter serotonin and is not a transporter, either.

The geneticist Tim Newman, a biological anthropologist at NIAAA, claims that in order for these MAO A variants to be retained over the eons, they must confer some reproductive bonus for the chimps, gorillas, monkeys, and humans who carry them. What we see as dangerously out-of-step behavior, he said, could be merely out of context. "Bold, aggressive males might have been quicker to catch prey or detect threats." But MAO A didn't become dominant because, for the most part, if the bully ape was that vicious, like a young thug in the *Sopranos* series, he often got whacked before he could breed. Human social evolution may have required the proliferation of many kinds of emotional and cognitive capabilities, says Newman's NIAAA colleague David Goldman, who added that social development may have been enriched by variations in human impulsivity.

In an oft-cited 1993 study, Harm Brunner, a Dutch geneticist, found a genetic connection that could help explain the abnormal serotonin levels found in some criminals. In an extended Dutch family full of nasty, hyperaggressive males, Brunner found some men who completely lacked the MAO gene.[36] Later, experts found that men who carry an MAO A variant presumably have less circulating 5-HT, and display aggressive, impulsive, and violent tendencies. Women also inherit the MAO A variant, but they are seldom

VIOLENCE   201

studied, since the effects are simpler to follow in men, who have a single X chromosome.

Studies of twins, especially those separated at birth, support the heritability of aggression—of which estimates vary from 44 to 72 percent in adults. While no single gene encodes for aggression, perhaps variants in genes that regulate the activity of neurochemicals such as serotonin, like MAO A, may contribute to people's susceptibility to violent, impulsive behaviors. It's not hard to see these MAO A/aggression findings as running parallel to Raine's observations linking lower PFC metabolism and violent, antisocial predispositions. There is, however, no one-to-one causal relationship between serotonin and aggressive behavior. Whether aggression and violence will occur when serotonin deficits exist in a person's brain will depend on that individual's personality—the intensity of one's impulsivity—and his or her social world.

In a remarkable long-term study documenting the dance of genes and environment, an international team studied the effects of violence and abuse on multigenerational families, and made several major discoveries about MAO A. Scientists led by Terrie Moffitt of King's College London and the University of Wisconsin, Avshalom Caspi, also of Wisconsin, and colleagues from the University of Otago in New Zealand studied 1,037 children, 442 of them boys, born in Dunedin, New Zealand, in 1972 and 1973. Tracking the children as they grew from age three to twenty-six, the scientists examined their genetic makeup and how they were raised. By age eleven, 36 percent (154) of the boys had been maltreated, and 33 severely. Some of the boys continued to act abusively and violently as adults, but most did not. A single gene helped explain why only a fraction of the mistreated boys became violent aggressive adults.[37]

The team found that 85 percent of the severely maltreated boys who had the weakened MAO A short variant manifested antisocial behavior. They committed robberies, rapes, assault, and persisted in fighting, lying, and stealing; they showed a lack of empathy or remorse for their actions. These abused children—12 percent of the whole group—accounted for 44 percent of the violent-crime convictions among the group. If abused boys had the shortened gene that caused their brains to underproduce MAO, they were nine times more likely to become antisocial. Abused boys with normal MAO A gene activity, on the other hand, were no more likely to exhibit impulsive behavior or commit crimes than those who grew up in a healthy family.

Simply having the low-MAO variant did not guarantee that a boy would go on to a life of crime and personal anarchy. Indeed, one-third of the human race has this gene. Unrelated to social class or ethnic group, but randomly distributed throughout the world, it is much too common to be of any use in screening or predicting who might become violent. In the Dunedin study, MAO A's link to aggression only emerged when the child had been abused. Thus Moffitt suggests, "the best strategy for preventing violence is to prevent child abuse." The gene's effect was more difficult to study in the girls with their dual X chromosomes. The fact that one protective gene variant could cancel out the "bad" one on the other X chromosome might help explain why females in general are less prone to criminal and severe antisocial behavior.

The Dunedin group also saw that another polymorphism, a high-activity allele within the MAO A gene, conferred a veritable gift upon its bearer. Boys who had been abused but had higher than normal levels of MAO A were unlikely to become aggressive-impulsive problem cases as teenagers or adults. Two-thirds of humans carry this high-activity genotype. The high-octane MAO gene confers the advantage that the bearer's serotonin system may be more efficient, more quickly returning to a balanced set point after, say, a stress-induced imbalance. The low-MAO gene carrier, theoretically, might suffer more from a traumatic event and have more trouble recovering after it.

In the abused children, carrying the high-MAO gene allele might have conferred upon them a "trauma resistance," a psychotropic buffer against the storms of their childhoods. One might imagine that this extra boost of the neurochemical-regulating enzyme, in turn, enabled these kids' OFC/ventromedial circuitry to better handle the emotions and impulses that no doubt roiled their minds. Before the Dunedin study, says Bruce Price, "we used to say that, for some reason unknown to us, some people are a hell of a lot more resilient than others. Now, resiliency possibly becomes biochemistry." (In terms of the remarkable difference MAO gene variants appear to play in the lives of the mistreated New Zealand children, one might suspect that, as in David Zald's study of temperamentally more agitated men with higher resting ventromedial brain activity, there is an MAO A effect at work in personality. Indeed, Zald is investigating the serotonin system.)

The PFC in a Court of Law

So we return to the big question: do violent people bear personal, even legal, responsibility for ruffling society's tranquillity if their brain physiology, neurotransmitter systems, or genes operate differently? "There is a link between an increased risk of violent or aggressive behavior and PFC deficits—no question about it!" Monty Brower exclaims. "And as a forensic psychiatrist, I need to be alert to the potential medical and legal significance of frontal lobe function when I'm assessing anybody. There've been individual cases where I had no problem saying to the court that a frontal lobe lesion or dysfunction substantially accounted for a particular violent or criminal act, or was a major contributor to it."

But, he continues, there is no fixed set of diagnostic criteria medical experts can cite when asked by the courts to assess mental illness, brain damage, and behaviors involving a person in violence or crime. "When I hear someone is aggressive, I want to know exactly what did he do, to whom, with what, under what circumstances, with what prior history? And what was he thinking? To simply label an action as violent and aggressive doesn't carry much information." For the forensic psychiatrist, the dim distinctions are a major defect in the heaps of academic data. "So you end up," he says, "with 'experts' attributing violent behavior to some problem in prefrontal cortex, but neglecting to note that the person was intoxicated at the time of the crime, or had an extensive history of substance abuse, or a history of violent behavior that was controlled, rather than impulsive, in nature."

There is, however, clearly an increasing trend for defense attorneys to make "mitigating factors" courtroom assertions: that violence and criminal behavior is caused by frontal lobe dysfunction. And, as some neuroscientists gossiped, the tactic thrived especially in the area around Irvine, California, where Adrian Raine and Monte Buchsbaum initially did their neuroimaging work. For a time, the UC Irvine PET-scanning center generated lots of business in the courts of Southern California. Brower, too, notes that "in some states with the death penalty, brain scans are turning up as a regular feature of sentencing hearings in capital cases." Although attorneys are willing to run with the argument that there is a general association between crime and frontal lobe dysfunction, and "therefore it excuses my client from wrongdoing," in Brower's opinion it is not valid. "Courts should not give these arguments credence," he declares.

"Yet I have no objection, in principle, to an attorney or court saying, 'Gee, this person seems to have impulsivity and self-control problems. Is there something wrong with his frontal lobes that helps explain, understand, or enables us to more appropriately respond to his behavior?' Add environmental pressures of a bad childhood and other social issues to the mix, and it becomes increasingly difficult to assess 'wickedness,' or blame. A weakness of the scientific literature on brain dysfunction, and violent behavior generally, has been to say maybe these people are more prone to getting head injuries because they do silly things like drive cars while intoxicated. So when you look at these people and their history of behaviors, you're at a loss to say anything robust and coherent about where the frontal lobe problems come from in the first place—and how much it was cause or effect," Brower says.

It could be quite a circle: genetically conferred traits added to a childhood-traumatized brain leading to risky behavior and further head injuries compounded by drug-taking might result in cognitive "decrements." People with executive problems might well find themselves increasingly frustrated, picked on, thrust into antisocial behavior. "It isn't that unusual," says Brower, "especially if you have educational difficulties, don't get much reward from school, and your role models don't bother much with education and get what they want from life on the streets. You might well say, 'This is an easier way to go.' Are prisons full of people who seem to have these frontal lobe problems? Do frontal lobe problems explain why they're there? Or are 'frontal lobe problems' just another stigmata of the luckless, deprived, neglected, abusive backgrounds that afflict persons who wind up in prison?"

The Harvard psychiatrist James Gilligan, the author of *Violence: Reflections on a National Epidemic*, has long studied the inner sources of extreme aggression, focusing on the dynamics of shame and humiliation, and how the persistent onslaught of such "soul-killing" emotions contributes to violence—in individuals and society.

Although Brower finds Gilligan's argument less persuasive on the societal level, when he sees patients in hospital wards with frontal lobe impairments who have been assaultive, shame and humiliation often turn out to be triggers. "These people perceived themselves as being shamed, humiliated, disrespected, so they become angry and respond the way they do when they're angry: they lash out," he says.

The capacity of someone with frontal lobe dysfunction to process and manage shame and humiliation is diminished. Their ability to screen all that out, to "relativize" it, put it in context, is defective. They're stimulus-bound, says Brower. "They say, 'Somebody is putting me down.' They don't say, 'Well, this happened yesterday, too. The guy's a jerk.' No, they're in the moment. They cannot say, 'Wait a minute, maybe I shouldn't do this because if I do, there's gonna be a consequence, so it isn't worth expressing my anger violently here and now.' They are unable to go through that mental process. That doesn't mean they're immune to the human experiences that provoke aggressive behavior, but that their ability to process, subordinate, and inhibit it is not the same as someone whose prefrontal cortices are intact."

To illustrate what it's like to be a patient adrift in the nebulous diagnostic waters of the forensic psychiatric world, Brower offers the example of "Mike," a kind of fictional composite whose experience typifies that of some patients he has encountered. As a young man, Mike ends up in a maximum-security forensic hospital. He has already been examined by many experts and categorized as an irredeemable sociopath. Mike's history reveals him "to be impulsive and distractible in the way people with attention deficit often are. Although ADHD was never diagnosed," Brower says, "it's reasonable to hypothesize it is a factor in his childhood. He grows up in a neglectful family, in which he is subjected to sexual molestation by a neighbor."

By age seven, Mike has begun a hegira of serial institutionalization in one kind of youth facility or another. He is both socially and sexually aggressive. He also sets a couple of fires. "Yet," Brower continues, "he appears quite intelligent by neuropsychological measures." As an adolescent he outgrows the Department of Social Services care and not surprisingly accumulates a juvenile record—not a terribly violent one, but one including substance abuse, petty theft, and low-level street crime.

Then, at twenty-one, Mike is hit by a truck. He suffers a severe head injury, undergoes neurosurgery, winds up in a coma for weeks, and is nearly given up for dead. But he makes a remarkable recovery, goes into rehab, and is released, essentially, back to the streets. It is then that he begins showing up regularly in emergency rooms and at psychiatric hospitals in varying stages of crisis. "He is threatening harm to himself, very agitated, out of control," Brower recounts. "But nobody can figure out what is wrong. His records list a raft of diagnoses—psychotic, schizo-affective, bipolar, impulse control disorder, substance abuse, or antisocial personality disorder." The assessors are clueless.

"After one too many assaults in another hospital," explains Brower, "our prototypical patient Mike ends up in court and gets committed to a maximum-security forensic hospital like mine. As soon as I hear, 'There's someone who just been admitted and he's impulsive, disinhibited, and has some history of head injury,' the thought immediately pops into my head, 'Well, sounds like a guy who's got a problem with his orbitofrontal cortex.' So, let's say I go and find Mike's records at the hospital where he was originally treated for his head injury. I look at the CAT scan, and voilà—Mike clearly has extensive damage to the anterior temporal poles and frontal lobes, particularly the orbitofrontal PFC." Such injuries, Brower says, would be typical of head trauma from a motor vehicle encounter like Mike's. Obviously, before his injury, our fictional patient was well on his way to a life of crime. But no one has seemed to notice that Mike's coma-inducing injury might have something to do with his later behavior. Instead, "The experts conclude Mike is predominantly an antisocial personality, and that is ninety percent of his problem. So the forensic psychiatrists decide the guy is better dealt with by the criminal justice than by the mental health system."

This misdirected diagnosis speaks to a rather astonishing lack of knowledge in forensic clinical and medical circles of neuropsychiatric disorders and brain injuries. Well, Brower replies, prefrontal injury is commonly misdiagnosed. Ironically, he adds, after the injury a patient like Mike was "not very good at being bad anymore. With the loss of his capacities for foresight and effective planning, and his increased tendency to be stimulus-bound, disinhibited, and drawn to whatever was immediately exciting in his environment, he wasn't adept at devising and carrying out antisocial acts. Although he still had many of the same antisocial motivations, they now played out in a much shorter-term, spontaneous improvisational, in some fashion benign way."

So Mike might steal something from another patient, and when it is patently obvious the object is someone else's, he still tries to pass it off as his own. "He saw it, loved it, took it," Brower says. "Then somebody asks him about it, so he has to come up with an explanation. There is a sequence of impulsive decisions—none of which is carried out in anticipation of other decisions or consequences that might occur down the line. Mike lives in a temporally foreshortened world where what he said twenty minutes ago has little or no relevance to what he says twenty minutes later. This is not necessarily because he forgets, but because the part of his brain involved in keeping track of the 'I need to maintain consistency in my behavior in the future based on what I said in the past, and twenty minutes from now I need to be behaving consistently with what I say now' is gone. Gone is the ability to see the memory of what he said as having significance for his conduct in the present or future, and being able to compare these things." One remembers Joaquin Fuster's notion of PFC function as "the bridge over the river of time" to see what Mike is missing.

Mike's labile temperament, his rapidly changing emotion in response to minor provocation, is also related to his head injury. Labile temperament? "Think of the wind blowing through a field of grass," Brower offers. "First it goes one way, then it goes the other. Swish, swish, swish . . . One minute he'll be laughing and joking with you, and five minutes later something will perturb him, maybe or maybe not related to you, and he'll grow angry, confrontational, and belligerent. Ten minutes after that he'll be joking again." Even in a highly structured environment, patients like Mike constantly become involved in minor disciplinary infractions that escalate into trouble when they are confronted. Since stimulation causes someone like Mike to become emotionally overloaded and lose self-control, how is he going to move to less restrictive levels of care, and eventually get out of the hospital?

Ultimately, someone like Mike might able to live in a structured residence, where the staff understands the nature of the injury and can mobilize specific therapeutic interventions to manage it. But it's a long road on which Mike must pass through successive layers of standard neurocognitive rehabilitation. He has two things going against him: crime and brain injury. Most state systems of forensic mental health care don't have the resources to provide the right treatment for patients where brain injury is a prominent part of their difficulties. Rarely are neurobehavioral or neuropsychiatric units specifically designed to treat the criminal-patient. Up to 70 percent of the people in correctional mental health settings have varying degrees of both brain injury and cognitive defects. And this often goes unrecognized by the system.

Brower has a theory about treatment for antisocial disorder patients with PFC defects, based on evidence that suggests that the anterior cingulate cortex is essential for processing feelings for others. Attachment theory, developed by the psychologist John Bowlby in the 1960s and 1970s, proposes that reciprocity in life's first relationships is the foundation of healthy development, for mammalian life in general and humans in particular. The baby's body language—reaching out, clinging, smiling—is reciprocated by the adult's, especially the mother's, holding, soothing, and smiling. These responses strengthen emotional bonding and give the baby a feeling of well-being. This positive sense of security may be the primary goal of attachment, and thus it is an early regulator of emotional experience. Further, it may lay the groundwork for Theory of Mind and social consciousness, and its disruption may be at the root of some forms of mental disorder.[38]

The critical period of anterior cingulated–driven limbic development may coincide with what Bowlby called the "attachment-in-the-making" phase of life. One attachment theoretician suggests that in the second quarter of a baby's first year a neural control system begins to develop, perhaps the maturing ACC, that begins to monitor the "earlier amygdala-dominated limbic" emotional system. Being starved of attachment during this first year could have long-term effects on the structure and function of this ACC/limbic network.[39] The ACC is a key structure, as we've said, linking the limbic to the PFC, especially the orbitofrontal cortex so vital to social and emotional processing. These circuits are involved in appraising and assigning value to events in the outside world and to memories and other internal constructs of the mind.

The violent aggressor, Brower thinks, may have some kind of ACC dysfunction relating to attachment. This will be the most severe for psychopaths. Brower explains, "People who in some biologically deep way seem to lack the ability to identify, empathize, have a mutualistic experience with another human, and who are therefore prone remorselessly and without conscience to exploit others for their own gratification, may have a fundamental defect in their development, and consequently in their adult capacity to form attachments. Maybe there is something wrong with their ACC, or an environmental input that impairs development of ACC, which has its expression in the neural substrate of behavior in ACC or both."

Psychopathic characters often obtain minimal success from conventional talking therapies because, Brower thinks, "it's not on their agenda, either internally or socially, to form attachments with others." This kind of severely antisocial patient finds the close approach intolerable; any attempt, however faint, to form a bond or alliance may in fact agitate him. He will try to flee from it, or indulge in some acting-out behavior rather than try to form an effective attachment. Perhaps these people are beyond help. Brower is prepared to concede some people are constitutionally, and by experience, simply not able to form and benefit from attachments. And many of those people are in the prisons. They're the likely recidivists, the recurrently violent who may be condemned to persist in antisocial and exploitative ways of life. "I'm not a Pollyanna, saying this works for everybody. In fact, for some, attempting it will probably not be effective. Instead, what they may need is ten years' time behind four walls."

What concerns Brower about the reductionist framing of the argument about the PFC, though, is that it suggests there are people for whom there is no way in. But in the right structured rehab environment, he believes, a person with frontal lobe defects that result in impulse control problems can be helped. "If his ACC is okay, he can form attachments, and use them to get better. He may never have had a satisfactory experience of obtaining it, and, indeed, may become quite discordant, confused, and agitated when they seek it. If the therapist can establish a relationship, then it can provide a context for whatever other treatment—whether it's medicine, cognitive behavior therapy, group therapy. Others find these violent people's behavior distressing, their nature so obnoxious that quite understandably they choose not to sit down and try to find a way in. It just happens to be my interest, my motivation, my fascination that leads me to be interested in finding a way in with these kinds of patients."

All humans are capable of mayhem and aggression. "One cannot attempt to understand the violence and criminal behavior of others unless he is willing to acknowledge at some level that 'there but for the grace of God go I'" says Brower. "The basis for empathy is being able to find a piece of one's own experience that one can relate to in the experience of the other and see as shared. If I'm with someone who's committed a terrible, violent crime, and I'm utterly intent on denying that this behavior could ever be part of anyone like me, then how can I ever talk or work with that person?"

More cynical or realistic, perhaps, is Dorothy Otnow Lewis. How is it we pour millions of dollars into books about Ted Bundy, Jeffrey Dahmer, and the like, she asks, but are nevertheless so willing to sacrifice deeper knowledge about them in the interest of doing away with them? "Maybe Ted Bundy was right. I suspect we are all far more curious about what the murderer did—the gory details of the crime—than about why he did it. It's the act of murder that fascinates us and tickles our own limbic systems. No wonder people fight for seats at executions."[40]

In the end, Brower and his colleagues are walking a tightrope between those who declaim that all these people are the spawn of Satan and should be locked up forever, and those who demand that since they are brain-damaged they should be let out. "That's what makes this work so interesting," Brower responds. "We have the opportunity to grapple with these issues and be struck with the wonder and complexity of human beings. We can appreciate the importance of individual dignity, free will, and moral choice, and also appreciate and understand the role of illness, the importance of compassion, and the ability to make relevant moral distinctions about behavior."

So inevitably we wonder what are the neural substrates of moral choice; what role does the PFC system play in making ethical distinctions and decisions about behavior? The glue of social organization is the ability to process information about self in the context of others, a calculus at which many antisocial personality disordered people seem deficient. Indeed, some social calculations require extremely complex and conflicting mental operations.

The In-Brain Ethics Committee:
Moral Values and the PFC

The game Diplomacy is an exercise in trust, the appearance of trust, and in persuasion, cunning, and deception. In the game, set in Europe on the eve of World War I, each player dons the mantle of a European empire. Through dealmaking and dealbreaking, and waging war, the power brokers vie for control of the world. Anyone who's enjoyed playing Diplomacy, or one of its many variants, knows how exhilarating is the forging and betraying of secret alliances on a global scale. The machinations of realpolitik constitute an intoxicating form of social cognition, a brew of cooperation, coercion, foresight, and chicanery—with a pinch of ethical concern and moral reproach.

One would prize the chance to have Machiavelli in the scanner, for no one has articulated the rules for playing the grand game as deftly as did he in the early sixteenth century. Not long ago, a reviewer rejected a paper on the brain bases of morality from the Brazilian researcher Jorge Moll, saying that it could never be a topic for neuroscience—that morality was the preserve of philosophy alone.[41] Wrong. Neuroscience is successfully peering into the brain to see how we wrestle with life-and-death issues such as abortion, the death penalty, and feeding tubes, the many social impasses in which the choices are structured in deep conflict. Such dilemmas can arise from short- and long-term goals, means versus ends, self-interest versus utilitarian goals more advantageous to society as a whole.

Related to deception, lying is a dark side of moral reasoning. The king of research on the neural substrates of lying and deception is Daniel Langleben at the University of Pennsylvania. At his behest, students played a standard laboratory card game known as the "guilty knowledge test," involving hiding a card, then denying you have it. From brain scans it was clear: lying takes extra work. Not surprisingly, the conflict-monitoring ACC is more active during lying than when one is telling the truth. Other brain areas, including the left prefrontal cortex, are busy suppressing the so-called prepotent, truth-telling response. The premotor area is also busy inhibiting "tells" and establishing the poker face and poker body. That dissembling takes more neural energy than truth-telling suggests to Langleben that truth is the brain's default state and must be overridden when we dissemble.[42]

At the opposite end of the moral spectrum from lying is cooperation and its righteous cousin, altruism. What are the brain substrates of these social calculations? Emory's James Rilling and colleagues scanned participants as they played Prisoner's Dilemma.[43] Here, two "criminal partners" were separately interrogated about the same crime. Not knowing what the other would say, each individual had to decide whether to rat out the other or stick together. Over twenty rounds, the partners pushed buttons to indicate whether to cooperate or betray. The rules were that if both kept silent, each would receive $2; if both squealed, each would get $1. And if one kept his or her mouth shut while the other defected to the authorities, the rat would get $3. Clearly, betrayal nets the biggest personal gain in the short term. As is usual in Prisoner's Dilemma, however, most players learned to resist betrayal, as long as the other partner did, even though both received a lesser reward for winning over the long stretch. How is this behavior, which has puzzled social scientists for so long, embedded in brain systems?

The scans revealed that during trials demonstrating loyalty, reward centers, including the nucleus accumbens and the PFC's orbital, ventromedial, and subgenual ACC (BA 25), glowed as if the participant had just won a prize, as if the brain were congratulating itself for loyalty. The social act of cooperation was paramount here; merely getting the money in this context didn't trigger the brain activity. Tellingly, when the participants were told they were playing against a computer, their brain reward centers mostly stayed quiet. And they were more willing to betray the machine program.

The reward circuit for cooperation with fellow humans is powerful enough at times, it seems, to override the allure of short-term self-interest. Anticipation of reward may be what engages the brain circuitry here; the preview of a future mutually beneficial relationship trumps immediate personal gains. Players did not remain loyal, however, after their partners betrayed them. There, the most prudent thing to do is return the betrayal and take the money. There is probably no neural reward for aiding a betrayer. This is played out in countless other stories of tit-for-tat and payback, such as in stories about the Omerta oath of the Mafia and the fate of someone caught wearing a wire.

The adage "Revenge is a dish best served cold" has always seemed a less than accurate axiom. There are too many accounts of people driven wild by betrayal—the cuckolded husband, for instance—committing violent acts of revenge even in the face of severe punishment. So finding that the brain bases of vengeance contain a rewarding mental calculus isn't a great surprise. Swiss scientists observed that taking revenge against a perceived wrongdoer, even at a considerable cost to the avenger, engages the same brain network that blazes away when we experience social reward: the dorsal striatum's feel-good caudate nucleus (nucleus accumbens), plus the ventromedial PFC (BA 10/11), implicated in balancing reward/punishment costs and benefits. The same system, basically, as activated in Prisoner's Dilemma. Revenge anticipates a payback, the emotional satisfaction of which may trump the rational calculations of its cost.[44]

The revenge findings "chip yet another sliver from the rational model of economic man," notes Stanford's Brian Knutson. Revenge-seekers in the Swiss study exhibited at least two types of irrationality: they reacted via predominantly emotional pathways, and expended costly personal resources to ensure that the cheater "got what was coming to him." "The [Swiss] findings serve as a harbinger of future neuroeconomic studies," Knutson adds, that should try to reconfigure economic models to include neurobehavioral information. Imagine new economic theories that accommodate both "passionate" and "rational" forces, as well as delineating when and how they converge to influence personal choice.[45]

Another facet of moral reasoning involves judging one's role in an ethical dilemma. For much of the twentieth century, moral psychology has been dominated by rationalist models: that we use sophisticated forms of abstract reasoning to judge our personal agency in a moral act. In question here are not so much judgments about one's role per se but rather judgments about what's right or wrong, which may or may not be affected by one's sense of agency. A more recent trend, however, emphasizes the primacy of our rapid, emotion-based responses, which have deliberate reasoning tagging along ex post facto to buttress those intuitive decisions. Princeton's Joshua Greene, with Jon Cohen and others, wanted to disentangle the neural substrates of moral decision-making, and also see which of the psychological models best reflect these brain computations. Based on their imaging observations, Greene and company colleagues now offer a synthesis of these two models.

The Princeton team argues that "personal" moral judgments are driven predominately by emotional processes and are up close and personal, while "impersonal" ones are impelled more by rational, higher-level cognition, and thus are more emotionally distant. There are possible evolutionary rationales for this distinction, Greene says, the evidence of which we can glimpse in the social conduct of great apes. The Emory primatologist Frans de Waal and others have long observed complex emotional interactions among chimps, suggesting that our common ancestors lived intensely social lives "guided by empathy, anger, gratitude, jealousy, joy, love and a sense of fairness." And all in the apparent absence of moral reasoning. Since humans, on the other hand, possess a higher capacity for sophisticated reasoning, it makes no sense if this faculty were absent from our moral judgment.[46]

So how does the brain resolve this tension between social-emotional, gut decision-making and the rational calculus required to make wrenchingly painful adjudications that are morally appropriate? Greene's group first scanned volunteers while they pondered dozens of moral conundrums including an off-the-shelf philosophical dilemma known as the Trolley dilemma. The first version of the problem is thus: to save five passengers who will be killed by a runaway trolley if it continues tearing down the track, you can flip a switch to reroute the train, but in so doing you will crush a passenger standing on the alternate track. Should you kill one person to save five? Most people say yes.

The Footbridge dilemma presents a variant scenario: you and a stranger are standing on a footbridge spanning the tracks, and the only way to save the five people from dying is to push this bystander onto the tracks. Should you? The majority of people say no, even though the net result is the same as in the Trolley dilemma. While the Trolley scenario included a moral quandary, the dilemma was considered impersonal; thanks to the distancing switch mechanism, you are once removed from the agency of killing another person. But the Footbridge version is intimate. Other personal moral problems involved whether to steal an individual's organs to distribute to five other people, and the triage-type impasse of who to throw off a sinking lifeboat. Other impersonal dilemmas included such conflicts as keeping money found in a lost wallet or not. As controls, the subjects also weighed "nonmoral" puzzles, such as whether to travel by bus or train given certain time constraints, or which coupons to use at a store.

There were differences in the way the brain fired in personal and impersonal moral evaluations. While subjects squirmed on the horns of a personal dilemma, such as the Footbridge's to push or not to push, the medial PFC (BA 9/10), posterior cingulate gyrus (BA 31), and junction of the temporal and parietal lobes glowed far more than during the impersonal or nonmoral quandaries. Executive, working-memory areas, such as the dorsolateral PFC, were quieter during the personal judgments, suggesting that emotional calculations can outweigh rationality in determining a solution to an intimate moral problem. Conversely, the impersonal and nonmoral dilemmas, while engaging abstract reasoning and working memory, activated dorsolateral PFC sectors more than the emotional areas.

Some participants fought off their emotional revulsion in the Footbridge dilemma and opted to push the stranger onto the tracks—the "utilitarian" response. They judged painful moral violations to be acceptable when these violations served a greater good. Participants opting for the utilitarian response hesitated sometimes as long as twenty seconds before they chose the horrific, but numerically appropriate, action. In general, when people chose in favor of committing a personal harm in the name of a greater good, they took longer to make their decision, suggesting that they required extra time to overcome an emotional response pointing in the other direction.[47]

No comparable pattern in people's reaction times appeared when they responded to impersonal moral dilemmas like the original Trolley dilemma, in which the train can be rerouted to save the five at the expense of one. The solution to this dilemma did not hold the same depth of moral violation as the Footbridge solution. The differences in reaction times and brain areas that lit up demonstrate that emotional-brain activity is not just incidental but exerts a direct force on moral reasoning and judgment—that emotions do lie at the heart of moral reasoning and can crucially influence one's judgments and choices.[48]

Personal moral violations, it seems, elicit powerful negative feelings that drive people to conclude that such rulings as the

Footbridge solution are unacceptable, even if they have utilitarian value. So to decide your deeply felt moral violation to be nonetheless the "right" choice, you must override your prepotent emotional responses with the iron hand of cognitive control—deploying that prefrontal genius for guiding reason and actions in accordance with goals and intentions in the face of competing pressures. When a subject responded in a utilitarian manner to the Footbridge dilemma, the significant time delay in his response reflected not only the involvement of abstract reasoning but also the engagement of control processes to suppress the emotional revulsion evoked by contemplating the idea of pushing someone to his death.

Since conflict was rife here, would the dorsolateral PFC also show increased firing during these neural computations, reflecting the engagement of abstract reasoning and cognitive control when one judged odious moral violations to be appropriate? That is, would increased dorsolateral PFC activity correlate with utilitarian judgment? And would the conflict-sensitive ACC be hopping as well? To find out, Greene focused on a class of dilemmas that brought cognitive and emotional factors into a more balanced tension. Such a crisis was the Crying Baby dilemma, which goes thusly: the enemy has overwhelmed your town, with orders to kill everyone. You and other townspeople are hiding in a basement. But as you hear the soldiers approaching your hideout, your baby begins to howl. As you cover his mouth, you realize if you remove your hand, his cries will attract the soldiers, and they will kill you all. But if you keep his mouth covered, the baby cannot breathe. To save everyone, you must smother your baby. Is it "appropriate" to sacrifice him?

As a control, the investigators provided an "easy" personal dilemma, in which a teenage mother must decide whether to kill her illegitimate newborn baby. Answer: of course not. This answer required little response time and little cognitive control. Here there was no greater good in the outcome to something otherwise deemed intensely immoral.

In response to the Crying Baby dilemma, subjects answered slowly and exhibited no consensus in their answers. As before, when they grappled with personal dilemmas, the medial PFC (BA 9/10), posterior cingulate (BA 31/7), and parietal areas (BA 39) fired away. Difficult personal moral dilemmas also engaged higher prefrontal

regions: the anterior dorsolateral PFC (BA 10/46) in both hemispheres. And, yes, the ever-alert conflict-monitoring anterior cingulate. In difficult moral dilemmas, a pragmatic cost-benefit analysis
was at the heart of the judgment. Reaching the "best" decision
required abstract, utilitarian "neuroeconomic" reasoning, plus the
cognitive control necessary to override the emotional brain that was
screaming against this violation.

This firing pattern was striking, because it seemed to contradict
the previous observation: that in personal moral judgment, the
dorsolateral PFC was quiet. Also interesting was the blazing of the
posterior cingulate —despite the fact that the PCC is supposed to be
an emotional processing unit. That the PCC is buzzing away during
such "logical" processing suggests that at no time during "rational
thinking" is the emotional brain totally silenced. The anterior
insula, also an emotional brain zone, especially tuned to disgust, was
also engaged. The anterior insula, as we've discussed, is associated
with risky decision-making in gambling and with rejection of unfair
offers in the Ultimatum game. Perhaps the insula firing here is
linked to the disgusted-with-reality aspects of difficult moral judgment, where we find ourselves gritting our teeth while endorsing
actions that are morally repugnant. In the teen-mother infanticide
case, where judgment is unanimous and swift, and there is little
temptation for a horrendously controversial choice, the insula is less
engaged.

Sectors of the medial PFC network, then, seem involved in judging moral dilemmas while simultaneously thinking about oneself.
The perimeter of BA 9 and 10, an area that may serve to fold emotional content into decision-making calculations, may be the zone
where emotional states, rationality, and self converge to construct
the mental representation of the dilemma and its final solution.
One should, however, distinguish the medial frontal gyrus from
medial PFC zones "behind" it, such as BA 8, which, says Greene, are
implicated in Theory of Mind. These areas have not yet been seen
to fire in unique patterns (as far as Greene and company can tell) in
specifically moral processing. Moral operations are part of social
thinking, but social-brain operations—such as Theory of Mind—are
not necessarily moral. While all emotions may at times contribute to
moral decision-making, certain ones, such as compassion, guilt, and
anger, are more central to our moral lives than others.

Greene's theory, then, incorporates both rational and affective psychological models of moral judgment. Resolving wrenching moral dilemmas marshals high-level reasoning and cognitive control processes, even as emotional-intuitive processes wield powerful influences. Greene thus challenges the black-and-white view that pragmatic, utilitarian judgments are entirely allied with cognition, while nonutilitarian judgments are emotion's dominion. Like Jeremy Gray and others, he suspects that all thoughts and actions, whether driven by clear-eyed judgment or not, have emotional tenor.

Indeed, binary labels such as "cognitive" or "emotional" may be of questionable distinction and usefulness. "The distinction may be a matter of degree, and therefore lose its usefulness as we get into complex areas. Alternatively," Greene adds, "one might render the emotion/cognition distinction in terms of contrast between mental representations that have direct motivational force and representations that have no direct motivational force of their own, but that can be contingently connected to emotional states that do have such force, thus producing behavior that is both flexible and goal-directed. According to this view, the emotion/cognition distinction is real, but it is a matter of degree and not necessarily clear-cut."

More broadly, Greene notes that for two centuries, Western moral philosophy has been largely defined by a tension between these poles of pragmatism and absolutism. Utilitarians such as John Stuart Mill argued that morality is, or should be, a matter of promoting the greater good. Moral absolutists such as Kant have argued that certain moral lines ought never to be crossed, for any reason. We see this dichotomy playing out in our culture even now, for example, in the unyielding gulf between pro– and anti–individual choice positions in the abortion issue.

Greene's research, where his test dilemmas distilled these philosophical tensions to their essences, may help explain the persistence of the polarities. The clash between utilitarian and absolutist perspectives may indeed reflect a deeper, more fundamental conflict arising from the structure of the brain itself. "The social-emotional responses we've inherited from our primate ancestors," he says, molded and refined by cultural experiences, reinforce prohibitions central to Kantian, absolutist positions. The moral calculus defining utilitarianism, though, "is made possible by more recently evolved structures in the frontal lobes that support abstract thinking and high-level cognitive control." Should this neural explanation prove correct, it would have "the ironic implication that the Kantian, 'rationalist' approach to moral philosophy is, psychologically speaking, grounded not in principles of pure practical reason" but in a set of more primitive emotional responses that are "subsequently rationalized," says Greene.[49]

Ultimately, Greene concludes, there is no "moral center" in the brain. Every sector operating during moral thinking operates during nonmoral thinking. Were we to segregate moral decision-making from every thought not specific to moral judgment—emotional reactions, Theory of Mind, abstract reasoning, strategic planning—"there will almost certainly be nothing left," he concludes. Morality is probably not a discrete species of thought but belongs to a family of diverse emotional-rational processes.

5

CREATIVITY
Art as a Window into the Brain

A rtists are, in a sense, neurologists who unknowingly study the brain with techniques unique to them," says the British neurobiologist Semir Zeki. Like diamonds of the mind, do creative experiences obliquely refract, not the world outside, but innate neural operations? Artistic endeavor and its products, "art," Zeki is convinced, are essential avenues the brain uses to explore, expand, and transmit the reality of our mind. As a window into the brain, art thus functions to drive human evolution. In creating the work of art, the artist "evolves" his or her own brain to a new stage: in Cubism in visual art, for example; rock and roll, jazz, or the sonata form in music. The work of art as a gift gives us a powerful new agency for interpreting the chaos of life for now and in future terms, while it allows us to enter and describe an otherwise inaccessible arena of neural processing.

Creative enterprises also serve as pressure-release valves for volcanic aggressions that are the outcome of eons-old neurowiring patterns. "In its pages, canvasses and scores," Zeki writes,"Mozart's *Don Giovanni* sets to sublime music the life of a lecher and serial rapist who would find no respite in the courts. His doom, announced musically in the opening bars of the opera, is dictated largely by his biological constitution. He faces that biological destiny with courage and dignity, as do Racine's incestuous Phedre and Shakespeare's Coriolanus, who is constitutionally blighted by pride and arrogance."[1]

When neurobiologists grow sophisticated enough to study the brain substrates of art-making, Zeki believes, their findings will revolutionize social organization, including education, law, and politics. Today that is a fairly radical assertion. Consistent throughout speculation about the brain and artistic expression, though, is the notion that creativity—innate, boundless, energetic, and strange as it can be—is the fruit of minds represented on the far ends of the bell curve. The paradoxical marriage of "sickness," mental and otherwise, and creativity is enduring. In the biography *A Beautiful Mind*, John Nash skates around the mysterious interface between mathematical genius and madness. And how does the architecture of the PFC play into the topography of *The Magic Mountain*?

A few years ago, Arthur Shimamura, a UC Berkeley brain scientist who investigates the functional results of prefrontal brain trauma, wrote a kind of speculative "neural history" of the nineteenth-century photographer Eadweard Muybridge. This monograph dramatically juxtaposes ideas of conventional mental health, the abnormal brain, and creativity. Muybridge's story is especially noteworthy because he achieved his greatest artistic successes after suffering a severe head injury and undergoing notable personality transformation. Shimamura proposes that an accident damaging the photographer's ventromedial/orbitofrontal PFC was the catalyst for both his troubles in life—and his creative liberation.[2]

Shimamura recalls how the idea of investigating Muybridge came to him. As a passionate landscape photographer, he had long appreciated Muybridge's work. Muybridge was famed during his lifetime and since, for his revolutionary and scientific use of stop-action photography to capture animals, and later, in his *Running Man* series, the human, in motion. His most famous work was commissioned by the former California governor and railroad magnate Leland Stanford, who wanted empirical proof that when his race-horses were trotting at full speed they "flew" with all four hooves off the ground. Muybridge designed camera attachments featuring an ingenious shutter system and set up multiple cameras at intervals along a racetrack to photograph the horse in motion. "Muybridge affirmed Stanford's supposition," says Shimamura—and that of millions of horse race fans for whom the "flying" thoroughbred is a fusion of nature, dance, and poetry. Muybridge captured the essence of art itself—the moment when terrestrial, earthbound life is left behind and for an instant one flies.

Muybridge was also one of the finest landscape photographers of the West and the first professional to document the grandeur of Yosemite. It was in preparing for a Yosemite vacation with his sons, camera in hand, that Shimamura read "a little travel history" of the park. In describing Muybridge's photography, the book mentioned in passing that he had suffered a severe head injury, and included some newspaper excerpts about a notorious event in his life. "At that point," Shimamura recalls in his office, "it clicked to me that his story ran very similarly to patients I see. Most biographers of Muybridge had found him an eccentric character, but they never linked his personality, talent, or the sensational events in his life to brain trauma." Shimamura was uniquely qualified to do so. "You need to have a fairly sophisticated background in how brain injury affects behavior, and an appreciation of photography, to link these things together," he tells me. "It was quite a revelation to have my interest in photography link with my professional life. It was fun."

In 1860, the thirty-year-old Muybridge was hurled from an out-of-control stagecoach, struck his head on a boulder, and was knocked unconscious. It took him months to recover. Reading about the accident and Muybridge's transformation, Shimamura suspected OFC injury. Some "transient anomalies," such as double vision and loss of the senses of taste and smell, suggested damage to the orbitofrontal cortex and nearby nerve fiber tracts. Ironically, these neurological records would have long since vanished, were it not for a wealth of evidence preserved in legal transcripts.

For in 1874, in a fit of rage, Muybridge shot his wife's lover, the playboy Harry Larkyns, after discovering that the baby of the two-decades-younger woman was probably not his offspring but Larkyns's. By this time he was already world-celebrated, and his trial for the killing of Larkyns in Vallejo, California, was arguably the O.J. Simpson courtroom drama of its time. The verdict was the same: acquittal. The murder and the trial happened near Berkeley, and Shimamura quickly dug into the archives. "I took a trip to the Napa

Superior Court, obtained documents of the trial proceedings, and made a pilgrimage to the location where Larkyns was murdered."

As part of his insanity defense, Muybridge's friends and colleagues came forth to testify that before the blow to the head he had been good-natured and emotionally stable, a prosperous bookseller and agent for a British publisher with a conventional persona. After the accident, by accounts from the trial, the formerly genial businessman became irritable, peculiar, a risk-taker subject to powerful emotional eruptions. "I could scarcely recognize him," stated a colleague who'd known Muybridge for twenty-five years. In the end, Muybridge stubbornly refused the insanity plea and put himself at risk for hanging. Ignoring the judge's instructions, however, the jury acquitted anyway, on the grounds of justifiable homicide. They too would have shot the seducer under the circumstances. (Note the sympathetic calculus of revenge at work in the jury box.) Nonetheless, suspects Shimamura, Muybridge's brain damage and consequent emotional volatility contributed to the violence of his attack on Larkyns.

Damage to the lower prefrontal cortex is today a common consequence of traumatic head injuries, such as in car accidents. The ventromedial/OFC sector lies next to the sharp ridges around the skull's eye sockets, and shearing against these bony edges produces contusions in the OFC and injury to adjacent areas in the anterior temporal lobe. In his day job, which includes studying orbitofrontal inhibition systems, Shimamura had observed that patients with OFC damage often have heightened responses to emotional images, events, and memories. These patients, as we've seen, often cannot seem to suppress or regulate their emotional states and cannot resist the rush that comes with taking risks, as you may recall from the legions of gambling studies of OFC patients. To Shimamura, the accounts of Muybridge's emotional transformation were fairly textbook descriptions of the behavioral aftermath of ventromedial/OFC injury.

After emerging from a coma caused by the head injury, Muybridge returned to his birthplace in England to recuperate. His physician there recommended outdoor photography as a therapy, possibly, Shimamura speculates, to steer the volatile patient away from troublesome contact with other people. Muybridge took it up with zeal. Fueled by a concomitant desire for hazardous adventure, he was soon accepting assignments in the wilds of Central America and Alaska, photographing in situations that make today's into-the-wild "extreme challenge" exploits seem like Cub Scout outings. At the same time, he displayed obsessive-compulsive characteristics associated with OFC abnormalities, says Shimamura, noting the vast, fanatical quantity of the photos of animals in motion, "which appear on the pathological side."

But from this obsession grew remarkable art and inventions. Muybridge took over forty thousand photographs of horse, dog, deer, goat, seagull, and human subjects performing various kinds of actions. Two of his books, *Animals in Motion* (1899) and *The Human Figure in Motion* (1901), were highly popular and "enlightened both artists and scientists." Muybridge invented stop-action photography that documented time sequences of rapid motion frozen in time. He mounted sequences of his horse photos on a glass disk, employing a lantern and lens to project the series in rapid succession so viewers saw the horse in full trot. This was, in fact, one of the first motion picture projectors. He also invented a "sky shade" device to cover part of a camera lens during an exposure so that brighter parts of a scene would not be overexposed or washed out, a precursor to today's array of filters.

Shimamura proposes that it was Muybridge's OFC injury itself that sparked his creative spirit. That Muybridge executed the designs for these ingenious inventions after his brain injury demonstrated he had not lost the problem-solving functions of the dorsolateral and other higher-echelon PFC areas. Yet something had been unleashed, liberated, in areas involving emotional control processing and the evaluation of rewards. "One could suppose that uninhibited emotion could act to heighten one's creative expression," Shimamura ponders. The relationship of PFC function and the artistic impulse is a seldom-visited frontier in brain research. But Eadweard Muybridge's case offers a tantalizing instance wherein artistic and inventive genius seems to have accompanied a significant prefrontal makeover.

"Shutting off one's PFC from time to time," offers Shimamura, "may actually enhance someone's creativity. The ability to take metaphors and analogies and see connections where others don't may involve making less use of parts of your prefrontal cortex. You want things to be connected that normally aren't. The frontal lobes are very good at selecting and tying things you know together. If you let that loosen up a little bit, maybe you'll start making connections where you wouldn't normally thought of doing so."

"As an experience, madness is terrific," Virginia Woolf once claimed, " . . . and in its lava I still find most of the things I write about. It shoots out of one everything shaped, final, not in mere driblets as sanity does." Central to creativity is its transformational ability: to take extant measures of information and combine them anew in ways that grant greater awareness of reality, which in turn give birth to more new ideas and actions. Yet it seems that, more than most, intensely creative individuals walk a tightrope between highly productive lives and debilitating mental illness. They have a higher rate of mood disorders, especially bipolar, depression, and addiction. The roll call of gifted and mentally afflicted artists is astounding. Here are a few names from Kay Redfield Jamison's book on the artistic temperament and manic-depressive illness, *Touched with Fire*: writers including Hans Christian Andersen, Balzac, Faulkner, F. Scott Fitzgerald, Hemingway, Hesse, Mark Twain, Dickens, and Virginia Woolf; poets Baudelaire, Blake, Keats, Lowell, Plath; composers Beethoven, Mahler, Schumann, Kurt Cobain; visual artists van Gogh, Gauguin, Michelangelo, Pollack, Rothko, and many more in each category. All mad, mad, mad at one time or another.

How to tease out the neural substrates of special creativity? In 2003, University of Toronto researchers found that the brains of extremely creative people may be more sensitive to the incoming sensations and perceptions that less sensitive individuals tend to discard in a process called "latent inhibition." Normally, your brain unconsciously screens out objects, sounds, events, and ideas that experience has shown are irrelevant to your needs. Unusually creative individuals may have lower levels of this latent inhibition; they may "remain in contact with the extra information constantly streaming in from the environment," suggests the psychologist Jordan Peterson. "The normal person classifies an object, and then forgets about it. The creative person, by contrast, is always open to new possibilities."

Lower latent inhibition thresholds might contribute to varieties of original thinking, especially when combined with a superior working memory's genius for juggling multiple ideas at once. Indeed, without good executive skills, low latent inhibition could be a bad thing. "If you are open to new information, new ideas, you better be able to intelligently and carefully edit and choose," adds Peterson. "If you have fifty ideas, only two or three are likely to be good. You have to be able to discriminate or you'll get swamped." Perhaps a partnership of low levels of latent inhibition and exceptional mental flexibility might predispose a person to mental illness under some conditions, and creative accomplishment under others. When they administered tests of latent inhibition to Harvard undergrads, Peterson's team found that students who reported unusually high scores in a single area of creative achievement were seven times more likely to have low latent inhibition. "Creative people are not as good at learning to ignore things," Peterson remarks.[3]

Common to both extreme states of mood disorders and the creative thought process is dopamine. Amphetamine and cocaine, which potentiate dopamine activity, cause people's latent inhibition levels to fall. In rat studies, high dopamine function leads to more novelty-seeking and less attention devoted to previous learning. When our dopamine levels rise, we become more exploratory, more open to the environment, wherein "things get perceptually renovelized," as Peterson puts it, because the "inhibitory strength of old categories decreases." Peterson, like Shimamura, suspects that the PFC's associational networks "loosen up," so that new ideas can be triggered and new patterns perceived. This does not mean, the investigators hasten to add, that highly transient effects of drugs like speed and coke are the way to open the floodgates of perception. You still must exercise the myriad decisions involved to make the sculpture, compose the quartet, write the lyric, choreograph the dance, or work out the mathematical theorem. And the drugs won't help with those processes.[4]

Some of the strangest evidence linking disinhibition to brain systems and creativity comes from the study of a devastating neurological disease called frontotemporal dementia (FTD), in which atrophy of the PFC and temporal lobes can, in some patients, result in temporary but formidable artistic proficiency. The leader of FTD research, Bruce Miller of UC San Francisco, has charted the course of artful madness in some of his patients. One FTD patient, Ed, a thirty-three-year-old car stereo mechanic with a tenth-grade education, signed up for a brief art course for beginners after showing no prior interest in or talent for the visual arts. Ed started with simple still-life sketches, and as his technique developed, he progressed to portraiture, then to stunningly detailed drawings of churches and haciendas he recalled from childhood.[5] (Illness apparently disinhibited these memories; they moved from gray background to brilliant foreground.)

Perhaps Miller's patients always harbored artistic impulses, but these failed to manifest until illness appeared. Other patients include a onetime conservative stockbroker whose ten-year, flamboyant artistic period landed him some awards in local art shows. Painting with feverish detail, he'd sometimes take hours to complete single lines. Another patient quit her job as a business manager to paint depictions of Santa Claus on gourds; an ex–advertising executive wrote a novel. When that failed he, like Eadweard Muybridge, left the comfortable life to photograph in Central America. A sixty-eight-year-old man began to compose classical music. All pursued their art with single-minded passion until the dementia consumed their minds. But before the end, their families and physicians saw that they had initiated a new way of thinking, centered more in visual and aural cues than language or social cognition.[6]

Comparing healthy creative people to patients with mood disorders, Stanford researchers explored the links between manic depression and creativity. Connie Strong and colleagues administered a battery of mood, creativity, and personality tests to forty-eight patients with successfully treated bipolar disorder, twenty-five patients successfully treated for depression, and forty-seven healthy "noncreative" controls. There were also thirty-two people in a healthy "creative" group composed of Stanford graduate students enrolled in prestigious product design, creative writing, and fine arts programs. This crowd included enrollees in the Joint Program in Design in the Department of Mechanical Engineering, studio arts master's students, and Stegner Fellows in writing.[7] (Some previous Stegner Fellows include Ken Kesey, Edward Abbey, Robert Stone, Scott Turow, Lan Samantha Chang, and Robert Haas.)

These creative people, Strong found, shared more personality traits with the bipolar patients than with either healthy "normal" or depressed people. The bipolar patients, too, showed higher creativity levels than did either depressed or healthy noncreative groups. The creative and bipolar groups were more open to new ideas and experiences. Having a wider emotional broadband is the bipolar patient's advantage, notes Strong. "It isn't the only thing going on, but something gives people with manic depression an edge, and I think it's emotional range." Both creative and bipolar subjects were also more likely to be moody and neurotic than were the healthy noncreative controls. Bipolar people might also have creative tendencies because they see the world from shifting perspectives, with the same environment appearing divergently depending on whether they are feeling manic or depressed. This multiple, or parallax, worldview might allow bipolar patients to be more receptive to new experiences and strange ideas. The bipolar group, moreover, had high creativity scores despite the fact that they were on medication. Bipolar patients commonly eschew medication because they don't want to blunt their creativity. Strong's findings suggest medication does not do that.

Another affinity between bipolar mania—with its force field of exuberance—and creativity may be the subjective feeling itself. When a person is mildly hypomanic, suspects the Yale bipolar expert Hilary Blumberg, he or she may have the energy of conviction to accomplish his or her art. But when someone is truly manic, Blumberg cautions, he or she has gone past the point where it is adaptive. The romantic view of mania often forgets this destructive side of the disorder.

While the Stanford study provided a road map for exploring the creativity/mental-illness connection, the study was not an imaging experiment. Psychologists at Lund University in Sweden, however, scanned participants—two groups of twelve men, one that had scored high on a creativity test, the other low—while they performed tasks requiring varying degrees of mental ingenuity. One task was an easy word-naming test. In another, measuring simple creative ingenuity, a man was given a word, such as "brick," and asked to think of as many uses for it as he could—from building a house, say, to using it as a doorstop to hurling it through a window. The creative men's brain activity differed significantly from the other group's in both the anterior PFC and frontotemporal areas. They had increased or unchanged firing in these areas as they moved from the rote tasks to the ingenuity-testing one, while the low-creativity men showed mainly decreased activity.

On personality tests, however, the creative men had higher anxiety scores than low-creativity men. And on an intelligence test, low-creativity men scored higher on logic, inductive ability, and perceptual speed than did the creative men, while both groups scored equally well on verbal and spatial tests. In other words, higher intelligence did not automatically mean the guy was creative, nor did low creativity imply he was not smart or smarter.

What are the brain bases of the creative spirit, innate in most of us? The few researchers in pursuit of this knowledge tend to find some kind of hemispheric bias in the frontal lobes. For decades it has been a street legend that the "right brain" is the artistic brain, while the "left brain" handles logic and rational thought. The *Drawing on the Right Side of the Brain* book has been a perennial best seller, indeed generating a cottage industry of self-lateralizing, how-to-liberate-the-artist-within books. But as we've seen, virtually nothing is absolutely monolithic or lateral about hemispheric allocations of labor in the PFC. The same goes for the desire to make and behold art. It may be that each PFC hemisphere plays a complementary role.

Stanford's John Gabrieli and then postdoc Carol Seger observed ways in which the two hemispheres work together and separately to process novel information. "Although most of us get distressed when people talk as if the right or left hemisphere don't talk to each other, you know, 'the analytical left versus the creative right,' there is something to it," says Gabrieli. He and Seger, now at Colorado State University, were exploring the neural bases of how we interpret novel visual-spatial information and organize it into meaningful categories when they touched upon this "creative" interplay of the frontal lobes.[8]

We intuitively do this kind of interpreting all the time when we see something new and then come to recognize it; we can do this at sophisticated levels of complexity in operations requiring pattern-recognition expertise, such as radiologists diagnosing X-rays and military pilots identifying enemy aircraft. Visual learning involves a transition from seeing and retaining countless inputs to organizing such images into significant abstract categories. Seger and Gabrieli offered the example of the letter Q: you must learn that Q, **Q**, q, *q* are all instances of the same letter despite their particular visual variations. This idea parallels Earl Miller's work with monkeys' "dog and cat" categorization.

Previous studies had shown that when children are learning to read, they show a right-hemisphere bias that shifts to the left hemisphere as they become skilled. As the programs of reading are learned, information falls into organized patterns and the hemispheric emphasis shifts. Similarly, in musical performances, novices show right-hemisphere dominance, whereas expert musicians show a bias toward the left. This phenomenon suggests that the right hemisphere tends to specialize in processing specific objects, events, and information, while the left specializes in patterns abstracted across the spectrum of incoming information.

The right hemisphere, furthermore, makes memory associations about specific, individualized visual things more quickly and accurately than does the left. And especially pertinent to artistic thinking, the right hemisphere is more responsive to new visual experiences than the left. The left hemisphere, on the other hand, performs judgments about "prototypical," generalized, categorical examples of a visual concept more rapidly than the right: does the object "fit" the category of "dogness" or "catness"? This is Miller's special PFC neurons at work.

To identify brain systems involved in this transition from novice to skilled levels of "inspired performance for the ages," Seger and Gabrieli scanned twelve people while they learned to distinguish between two new visual concepts. The subjects were given two lists of forty-eight examples of "artworks"—lab samples loosely resembling details from abstract graphics from the work of hypothetical "Smith" or "Jones." They had to learn to distinguish between the "paintings" of these two "modern artists." Each time the participants were shown a detail, they would decide whether Smith or Jones had painted it, and would be given feedback about the correctness of their choice. At first they could only guess, but over time most learned to classify the works by identifying subtle "stylistic" patterns in each. The participants differed considerably in their growing expertise. Six became proficient; four never really got the hang of it.

What was going on inside the brain? First, when all the painting fragments were novel, the scans revealed neural fires blazing in the right prefrontal and inferior parietal areas. As the "art critics" began to detect which painting details expressed the "style" of Smith or Jones, the left superior parietal hemisphere joined in the action. The left PFC slowly increased in activity from the middle through the end of the test. And the left PFC was significantly more active in successful critics than in those who never could pick out whose pictorial details belonged with which painter. The results delineated a dynamic network wherein during the initial naive stage, information about the paintings could only be processed as unique and strange visual patterns, and right-hemisphere areas were ablaze.

The critics had to analyze the features of the paintings, differentiate between the emerging patterns, and formulate rules that applied to these features and patterns, then make an inference based on such rules. The left PFC came into play only as they attempted to master the rule base, the "conceptual knowledge" of the differing "aesthetic" styles. The successful application of these rules led to a shift to more left PFC activity. The right PFC and parietal cortices in both hemispheres are involved in visual reasoning, and engage in visual-spatial working memory. So it made sense that the right hemisphere—both frontal and parietal—would be a big player here, both in holding the holistic image of the painting online and scrutinizing its details. But the left dorsolateral PFC blazed away only in the people who showed high proficiency in discriminating Smith's work from Jones's. This is consistent with evidence that the left PFC deals in analytic problem-solving, in formal reasoning, and may be more specialized for classifying, abstracting, and rule formation. The individuals who performed well said they noticed patterns and made rules about particular features, such as a "staircase-like appearance" in one set of examples.

Right-hemisphere PFC activity persisted for the duration of the trials in both the poor and astute observers. This right-hemisphere buzzing appears related to the constant processing of information about the specifics of the individual painting's visual and spatial properties that remain distinct from the category learning of "Is it Smith or is it Jones?" Indeed, Seger said, the right-hemisphere PFC processes should be "insulated" from the categorical knowledge in order to maintain their focus on the specific colors, lines, and spaces of each painting. The left-hemisphere PFC fired as participants generated abstract categories about each artist's oeuvre that would apply to each painting. One can imagine the whole PFC network operating in parallel processing and feedback loops across the corpus callosum, as experts like Robert Sternberg have suggested: "Systems that operate in parallel allow for the simultaneous and independent consideration of competing, and even contradictory, hypotheses."[9]

Thus the left hemisphere exhibited a kind of dynamic rewiring capability, based on the experience of analyzing one painting after another, fueled by information supplied by the right hemisphere. The rapid "educability" of the left hemisphere, then, should be useful in enabling people to quickly propagate and apply new abstract categories, while the right hemisphere remains a stable system for dealing with specific richness of new and exotic details in the visual informationscape.

How would this work with words? Gabrieli and Seger's team looked at the right hemisphere's handling of unusual noun-verb relationships. Although the left PFC is the language-processing hemisphere in most people, the right hemisphere is involved in some lingual operations, such as understanding conversation and prosody (the stuff of poetry: versification, the rhythmic and intonational aspect of metrical structure). The Gab Lab used fMRI to see if the left hemisphere tends to process closely related words, while the right is dominant for more uncommon verbal relationships.[10]

Some of the participants were asked to think of the first verbs to come to mind; something, obviously, we do when we talk or write. They were to serve up verbs that commonly fit with nouns they saw flashing on a screen, such as balloon, cactus, canoe, clay, clothespin, comedian, cyanide, flute, lollipop, rocket, roof, ruler, trowel, truck, and wreath. If the noun was airplane, for instance, the verb was fly. A second test group was to fashion unconventional verb pairings. Say the noun was dish: a bizarre verb was spawn. Some of the nouns for the oddball-verb series included: ammonia, amoeba, bladder, cave, clod, idol, penguin, rouge, salary, Sherpa, violin, wizard, and yacht.

When the participants generated conventional noun-verb pairings, areas of the left PFC were active, which was consistent with what everyone believed about word processing in the PFC. But when people propagated strange mixes, there was extensive firing in the right PFC and the ACC. Although the left PFC continued to fire as before, the right hemisphere was newly recruited, especially the anterior PFC including the abstractive and self-involved BA 10. This discovery underscored the idea that the right PFC networks, too, can be involved in specifically verbal processing. The test-takers had to evaluate each oddball verb they dreamed up against an internally programmed set of criteria, asking themselves, "Is this verb weird enough to go with that noun?" Was the right prefrontal area 10 recruited because unusual verb generation requires greater self-reflection and self-monitoring of one's performance?

The right PFC is ace in operations coupling distant semantic entities, such as metaphoric relationships: Homer's "rosy-fingered dawn"; Wallace Stevens's "green freedom of a cockatoo." People with right-hemisphere damage have trouble "getting" metaphors. Thus if the right hemisphere plays a part in resolving lexical ambiguity, this may partially account for the difficulties that right-hemisphere-damaged patients also experience in comprehending discourse and conversation in general, which is often replete with ambiguity: "He said to her he might go there!" Right-hemisphere patients are also less able to suppress ambiguous word meanings than healthy people.

Since the ability to access and generate extraordinary relation-ships between things is the essence of creativity, hemispheric spe-cialization for uncommon noun-verb relationships points to duality and suppleness in human thought and language, whereupon the right PFC network is the gleaner of the "raw material" of an artistic event, and the left PFC network is the organizer and shape-sculptor of these resources. Seger tells me that one of her students, Gwen Schmidt, has found that the right hemisphere is recruited for unfa-miliar, novel metaphors ("Rain clouds are pregnant ghosts") whereas more conventional metaphors recruit the left hemisphere ("She has a heart of gold").

"The real issue in neurobiology is what is the basis of the divi-sion," says Vinod Goel, who has puzzled over the question of hemispheric lateralization in his joke studies and more recent inves-tigations into the neural bases of aesthetic taste. "People are saying pictorial versus linguistic, analytical versus synthetic—a whole series of these dichotomies. Sure, I think there is a hemispheric distinc-tion. And I was surprised, because when I started I didn't think there were such distinctions. But you get such dramatic differences in so many domains in MRI studies—positive mood in the right, words in the left. But then again, you get things that don't quite fit. So you say, well, maybe it's conceptual versus nonconceptual, struc-tural versus nonstructural. It's an unresolved issue. We're still mis-conceiving what the distinctions are. Picasso, a right-hemisphere man? There may be an element of truth in it, but it is a very tiny element. We need to reconceptualize the distinctions, sort them out."

At what age are the "aesthetic" patterns laid down in the brain? The neurobiological wiring for music, for example, is quite precocious. Even infants exhibit an unmistakable sense of musical discrimination. Even six- to eight-month-old babies are sophisticated at picking up musical detail. At McGill University, 140 infants participated in a study to determine whether babies can remember complex music.[11] Investigators Beatriz Ilari and Linda Polka found that the babies could remember hearing a classical piano piece, even after a two-week interruption. While past studies had established that babies are highly attentive to sounds, most experts recommended they be exposed to simple songs like "Mary Had a Little Lamb." But Ilari's findings challenge the assumption that infants are ill-equipped to handle complex music. For her tiny subjects, Ilari, a violinist, chose "Forlane" and "Prélude" from Maurice Ravel's suite for solo piano Le tombeau de Couperin. She selected these pieces because of their evocative harmonies, complicated rhythm, and elegant texture, and because they are hauntingly beautiful. "We also wanted little-known pieces that parents wouldn't know," she said, adding that the babies came from a cross section of society.

She gave the parents a CD containing either "Forlane" or "Prélude" and asked them to play it to their baby three times a day for ten consecutive days. Then, after two weeks of not hearing the piece, the babies were tested. They were seated on a parent's lap in a three-walled booth, while researchers played eight excerpts from the assigned Ravel piece, intermingled with eight of the unfamiliar one. The researchers mounted a red light on each side of the booth to the left and right of the baby. One light would blink to attract the baby's attention. Once the baby looked at the light, one musical excerpt would be played through a hidden speaker mounted behind the light. The excerpt would play until the baby turned his or her head away in disinterest. The babies were 20 to 30 percent more attentive to the piano piece they had heard: babies exposed to the "Forlane" preferred it to the "Prélude," and vice versa. The babies formed an impression of the music and were able to retain this impression over a two-week interval. What information had the babies retained from their exposure to the music? "Was it the melody, or some more global acoustic features of the music?" Ilari asks. "This will require further investigation."

Just what do we know about the PFC's role in processing music? The longtime investigator of brain and music Takashi Ohnishi at Tokyo National University has observed that the planum temporale (PT) in the left hemisphere is bigger in adult musicians. Exposure to and training in musical experience in childhood augments PT firepower, strengthening its connections to the left prefrontal cortex. Situated on the upper surface of the temporal lobe above and in front of your ear, the planum temporale includes the classical "speech center" of Wernicke (BA 22), which links to an auditory association area, which in turn projects to the "back" of the dorsolateral PFC (BA 9). This circuit is vital for learning the association between the pitch of a sound and its verbal or written label, say, C-sharp. These neural differences—both those one is lucky to be born with, such as perfect pitch, and those achieved by practice— may allow musicians to hear differently, and listen to music more analytically, than nonmusicians.

Ohnishi scanned the brain activity of musicians and nonmusicians as they listened to a piece of music. In musicians, the left PFC and the planum temporale were dominant when they listened to music, whereas the right auditory cortex fired more in nonmusicians' brains. The degree of activity in the planum temporale and the left PFC correlated well with the age at which the person began musical training. Nonmusicians react to music more viscerally and emotionally, and would probably show more activity in the autonomic, limbic, and orbitofrontal PFC areas than musicians. Performers need a steady heartbeat and emotional control, plus tremendous memory-motor hookups that nonmusicians don't need. (There is an oft-told tale about a noted music critic who visits the legendary violinist Jascha Heifetz in his dressing room after a dazzling performance at Carnegie Hall. "Jascha," he asks, "please tell me what was on your mind while you were playing that spectacularly virtuoso passage just before the cadenza of the first movement of the Tchaikovsky Concerto." "Oh," says Heifetz, smiling, "I was thinking of how delicious the lox and cream cheese on a bagel would taste at the Carnegie Deli after the concert." After all, what is music if not the bridge of thought and emotion over time?)

Ohnishi found left PFC and planum temporale activity to be

correlated with absolute pitch ability. People with such innate ability had significantly larger left PTs. Thus his study revealed distinct variances in this auditory-prefrontal network in musicians
based on both nature and nurture.[12] And it gets more complicated: melody and rhythm may have different hemispheric dominance, with the right hemisphere more sensitive to melody and
the left to rhythm. Children near age six show some brain specialization for rhythm and melody processing, but to a lesser extent
than adults do, suggesting that the hemispheric bias in processing
different aspects of music develops as we grow.[13]

Mapping the brain's musical geography, it's also becoming clearer
why we cringe when a pianist lays down a wrong note in a crashing
chord, or a singer reaches for a high C but lands on a C-flat. Our
brain wiring makes us anticipate hearing certain tones in sequence,
report Petr Janata and colleagues. In the face of the awesome diver-
sity of music we hear, the situations in which we hear it, and our
spectrum of visceral and emotional responses to it, it is likely that
brain units that process tonal contexts reside in regions already pre-
disposed to handling interactions between sensory, cognitive, and
emotional information. The PFC, of course, is such a nexus.[14]

Western tonal music, admits Janata, now at UC Davis, has a "certain allure" for a neurobiologist as a stimulus for probing the cognitive machinery of the human brain. This allure derives from the
fact that the brain has structures that specify distance relationships
among individual pitches, pitch classes (so-called chroma), pitch
combinations (chords), and keys. These interval-specifying structures are arrayed so that specific combinations of notes and chords
sound more coherent to us when they are heard in sequence. These
intervals in sound, which exist in a "hyperspatial dimension somewhat analogous to the concrete reality of keys on a piano," says
Janata, "shape our perceptions of music and allow us to notice
when a pianist strikes a wrong note." The intervals help govern patterns of expectation arising when we listen to music. And indeed,
these relationships may underlie our fulfillment when we respond
emotionally to it. That is, we are pleased and satisfied when the
sounds "land" on these "right" neurotopographical targets.

Tonality, then, is represented mapwise in the brain. When a person in a test situation rates how well each of twelve tones drawn
from the chromatic scale fits into the preceding tonal context

defined by a single chord, chord progression, or melody, the tester's rating depends on the relationship of each tone to this context. Tones that do not occur in the key are rated as fitting poorly, while tones that form part of the tonic triad, the defining chord of the key, are judged as fitting best. (The chromatic scale consists of twelve equally sized intervals, called semitones, into which an octave is divided. On a piano, for instance, you play a chromatic scale starting at middle C by striking adjacent keys until you reach the C either one octave above or below middle C.)

Previously, the PFC had been implicated as a manipulator and an evaluator of tonal information, but before Janata began his imaging studies, no one knew whether PFC regions were involved in mapping musical motion onto the brain's "tonality surface." How do we recognize a song played in a different key as, in fact, the same song? To identify tonality-tracking cortical areas, Janata scanned eight listeners, each of whom had at least twelve years' musical study, for three sessions, separated by about a week. In each session the participants heard a melody that systematically modulated through all twelve major and twelve minor keys. Participants were asked to pick out specific tones that violated "local tonality" (notes that sounded "wrong" because they were not struck in the key the music was in at that moment). They were also asked to detect notes played by a flutelike instrument instead of the clarinet that was standard for the music.

These melody-listening tasks consistently lit up several regions in the temporal, parietal, and frontal lobes, as well as components of the thalamus and the cerebellum. The most consistent firing was along the superior temporal gyrus, although the extent was greater on the right side, stretching from the planum temporale forward. Each musician's brain tracked the sounds in a slightly different way each time he or she listened to the music. This may be one reason why the same music played or heard at different times often elicits different feelings. The clearest and most persistent activity during the tonal tracking, however, arose in parts of the orbitofrontal cortex, and in the rostromedial PFC, just behind the center of the forehead in both hemispheres. The rostromedial PFC, the investigators suspected, was tracking tonality in the same way for different melodies, while determining which key renditions it found most agreeable or satisfying. Consistent blazing activity in this region in all listeners led the scientists to postulate the rostromedial PFC as a central site for a tonality map.

The medial PFC, Janata insists, is situated in an anatomically ideal zone for the functioning of a tonality map. In the monkey, for example, connections to it from the primary auditory cortex are widespread, compared to those from other sensory cortical areas. Janata also saw activity in other higher-order auditory areas during the experiment—especially in the right hemisphere. He thinks the rostromedial PFC not only assesses and responds to the general "rightness," or consonance, of the harmonic accompaniment to a melody, but actively holds online this topographic representation of tonal knowledge. The rostromedial PFC, then, through a kind of harmonic working memory, may be the brain processor that enables us to recognize a song played in different keys.

Janata's group also saw that this mapping of a specific key by specific neuron populations in the PFC changed dynamically from one occasion to the next. Unlike common visual objects that differ in their spatial features, musical keys are abstractions that share core properties. The intervals among pitches within a key are identical for each key, making it possible to accurately transpose musical themes from one key to another. The interplay of short- and long-term memory banks of tonal information may also aid in generating a map. This dynamic arrangement in the PFC may help couple the moment-to-moment perception of tonal space with thoughts, emotions, and feeling-like-dancing, all of which may circle back to the PFC maps and participate in shaping the patterns of neuron fire within the rostromedial PFC while one is listening to Beethoven's *Emperor* Concerto or Jimi Hendrix's "All Along the Watchtower."

On the most basic level, the areas in heterosexual men's brains that respond to the face of an attractive woman are the same as those reward networks that light up for food, cash, and recreational drugs—a network that includes the medial orbitofrontal cortex. For the straight guy, the simple act of looking at a beautiful woman's face is a reward in and of itself. However, this circuitry does not fire up when he views the face of an attractive man. Obviously, the perception of art holds something rewarding for the brain, but how do you get from mate-choice and food-driven pleasure bombs to delight in beholding a masterpiece of painting, music, literature, dance, drama, or film?

The nascent field of neuroaesthetics is devoted to understanding the neural correlates of artistic experience. Neuroaesthetes confront such issues as: Why is great art "timeless"? What does it mean when someone says, "I don't know much about art, but I know what I like"? The neuroaesthetes see art, and love of it, as a product of brain organization and as being subject to its laws.

The evolution of the human brain has been widely accepted as giving rise to the artistic impulse and the evaluation of beauty in nature and art. The capacity to appreciate and create beauty in the visual domain through the novel arrangement of such basic elements as color, form, and movement arose during human evolutionary lineage as a trait not shared with any other primate. Prehistoric art "represents features salient in the human mind as it explores the world," said the UCLA professor of communication studies Francis Steen.[15] Although the expansion of the prefrontal cortex has been linked to the appearance of such traits, no empirical evidence had documented its role in aesthetic perception. In 2004, the Spanish anthropologist Camilo Cela-Conde teamed with neuroscientists to determine whether the PFC is indeed central to the workings of our visual aesthetic faculties.[16]

His experiment deployed an imaging instrument, magnetoencephalography (MEG), that allows for superquick readings of successive brain events, a sensitivity handy for investigating visual stimulation. As their working hypothesis, the scientists used the theory of multistage integration of visual consciousness based on Semir Zeki's research at University College London. Zeki has long focused his research on the brain substrates of visual arts, and has written two books about it: *Inner Vision* and *La quete de l'essential*, which he coauthored with the late French painter Balthus.

Zeki's model is based on evidence that the visual brain consists of a number of parallel multistage processing systems, each of which performs a special operation, such as processing color, line, or motion. The PFC, of course, integrates visual working memory, spatial location, attention, long-term memory, and so on. Zeki and colleagues observed that when people perceived some essential aspects of art, they showed specific neural firing patterns in line position "color centers" in the V4 region of the visual cortex. The scientists concluded that this processing system is equivalent to a perceptual system. Artists, they said, would intuitively exploit this

neural processing system for color and form to promote aesthetic
sensation. Because Zeki and colleagues did not really seek to identify the brain correlates of the aesthetic judgment of visual perception, Cela-Conde decided to find out if the PFC did indeed deserve
its reputation as the zone critical for perceiving "beauty."

Using MEG, Cela-Conde's group sought to locate brain areas
that were ablaze while subjects viewed pictures they deemed beautiful or not. They recruited eight female neurobiology students,
none of whom had aesthetic training, and who fell more into the science geek mode than that of art school cool kids. The images each
was asked to evaluate were culled from an enormous collection of
classic, impressionist, postimpressionist, and abstract paintings;
included were photographs of landscapes, artifacts, and urban
scenes. The photographic archive came from various collections,
the book *Boring Postcards*, and photos taken by the scientists themselves. The investigators used as a guide styles selected from the collection Movements in Modern Art of the Tate Gallery, London,
adding European XVII and XVIII Centuries and American Popular Art pictures. Their objective, says Cela-Conde, was to offer the
women a broad range of styles to maximize their choice of aesthetic
judgment.

To make the choices as technically balanced as possible, the scientists subjected the pictures to three evaluatory criteria. First they
assessed each picture's complexity—if they were judged too simple,
they were thrown out. Then they adjusted each picture's color spectrum using Photoshop. They tweaked extremes of illumination and
shadow to highlight details. And finally they measured the light
reflected by each picture, and the extremes were thrown out. In the
end they had assembled 320 pictures reasonably equal in complexity, color spectrum, luminosity, and light reflection.

While the women viewed the pictures, they raised a finger if they
considered the picture beautiful. Throughout the testing period,
the MEG data showed blazing activity in great portions of the cortex: the occipital, parietal, inferior, superior, and medial temporal
lobes (including the hippocampus); the dorsolateral, orbitofrontal,
and frontal pole (BA 10) areas of the PFC; and various motor areas.
But the left dorsolateral PFC fired only when the women perceived
beauty in the pictures upon which they gazed.

The MEG technology, furthermore, detected both the asynchrony of brain processing and just how long various processing subcomponents of the "aesthetic response" took. While visual perception was registered in about 130 milliseconds after the picture flashed on the screen, the PFC's "beauty recognition" occurred between 400 and 1000 milliseconds after the image appeared. The left hemisphere was generally more active, which the investigators interpreted as reflecting the ongoing decisions (beautiful or not beautiful) the women were making about every single picture.

The lack of a precise definition of "aesthetic perception" presented something of an obstacle for the experiment. For centuries the term has signified a sensory perception, but Cela-Conde and colleagues agreed that aesthetic perception must go further to embrace complex abstract factors. Visual art, for instance, is generally understood as involving the creation of objects produced with the intent of eliciting the experience of beauty. The apparent divergence between art and beauty is clear when we realize that some avant-garde art rarely arouses the sense of beauty per se in the public at large.

Since antiquity, artists and critics alike have rejected beauty as an essential ingredient of art, claiming that emotions far beyond hedonistic pleasure are the foundation of the aesthetic experience. Certainly Aristotle pondered the question in his thoughts on the nature of tragedy in drama. Cela-Conde, nonetheless, thinks it is hard to deny that people who visit art galleries and museums, read poetry, and go to the theater and dance do so in a quest for something approaching beauty.

So the researchers confronted a problem: how could the constellation of aesthetic experiences be reduced to simple variables? Cela-Conde cites a theory of a "general factor" or "basic knowledge" of arts proposed by the psychologist Hans Eysenck. And several studies have confirmed the presence of a "basic knowledge" of art in people with little or no formal training in the arts or aesthetics. Even though many factors affect artistic judgment—social, cultural, historical, biological, educational, and personality—and will undoubtedly affect the results of such a test, the scientists were convinced the participants had a sense of their own valuation of "beauty" that was different from "pleasant," "original," or "interesting."

The results do not imply a "brain center for aesthetics," Cela-Conde insists, in the sense that there is one location to which information streaming from all sensory channels converges. The visual system lacks this kind of organization, for one thing. But the dorsolateral PFC could be a crucial node in a neural network intrinsically dedicated to conscious aesthetic perception. Future research is needed to explore how experiences like art training modify visual aesthetic perception, as it may auditory perception in musical training; how aesthetic perception is different when one perceives beauty in nature versus in art, or in realistic versus abstract representation. It will also be exciting to discover how and when subcortical systems—such as the limbic and thalamic-striatal—are recruited to engage the entire thinking-feeling-body experience of beauty.

For Semir Zeki, the disturbing, arousing richness of the artistic experience, born from the subjective world of the individual artist, is based on a commonality in human experience that allows us "to communicate about art through art." This commonality, he proposes, is rooted in the brain's structure/function relationships. Zeki, the neurobiologist of the brain's visual anatomy and physiology, proposes that painters themselves have studied the perceptual system. Years before discovery of the visual cortex's orientation-selective cells, which respond only to straight lines and are considered the building blocks of form perception, Mondrian, for instance, in search of "the constant truths concerning forms," settled on the straight line as the basic feature of his compositions, notes Zeki.

Similarly, great painting seeks to distill on canvas the essential qualities of objects and surfaces as they change in time. Thus, says Zeki, a major function of art is to serve as an extension of brain function, namely to seek knowledge about the world and to create abstract ideas and patterns out of that knowledge—a prefrontal process. Abstraction is a crucial step in the efficient acquisition of knowledge, and without it, according to Zeki, the brain would be "enslaved to the particular." Art abstracts as well, and here again, in generating an object hitherto not present in the world, externalizes the inner mechanics of the brain. Abstraction, in Zeki's model, leads to a paradox: the refuge of ambiguity. The abstract "ideal" the brain labors to create from myriad particulars in its mental workshops can lead to a deep dissatisfaction in the "conscious mind."

Anyone reading Plato's *Republic* can sense on some level his futile struggle to build a complete world based on the Ideal, or for that matter, the frustration of someone trying to throw a ceramic vessel for the first or thousandth time, trying to achieve an impossible vision of perfection in the clay form. Michelangelo left three-fifths of his sculptures unfinished because, according to Giorgio Vasari, "the subliminity of his ideas lay beyond the reach of his hands." Many great pianists, too, have stated that two or three performances in a thousand approach realizing their inner-sound ideal of the music. Zeki would put it differently: "Michelangelo realized the hopelessness of translating into a single work or series of sculptures the synthetic ideals formed in his brain." The finished as well as the unfinished work, however, is blessed with the allure of ambiguity that allows the beholder to interpret the work in a multiplicity of valid ways. Thus art is a haven for unfulfilled ideas born and alive only in the brain through its abstractive processes.

In the work of Vermeer, Zeki talks about not the " vagueness or uncertainty found in dictionaries" but the certainty of many "different, and essential, conditions, each of which is equal to the others, all expressed in a single profound painting, profound because it is so faithfully representative of so much." Zeki reminds us of Dante's unrequited love for Beatrice, and that Wagner composed *Tristan und Isolde* as the "greatest monument to the greatest of all illusions, romantic love." But an illusion is a construct of the brain, Zeki adds. And the PFC, we know, is the supplier of workshops of illusions that we require to design and build any big idea.[17]

Musing upon Zeki's theories and others, Russell Epstein, at the Center for Cognitive Neuroscience at the University of Pennsylvania, has written an essay about Marcel Proust, metaphor, and the brain. Metaphor, that basic unit of prefrontal currency, enables the writer to relate one sensation, memory, or series of events to another, thus creating a common labyrinth of thoughts, remembrances, and inferences. One can then surpass a single sensation "in the focus of consciousness," says Epstein, and approach an awareness of the web of associations that accompanies this metaphor. Indeed, the process of elaboration by metaphor, some brain scientists think, underlies all thought.[18] By stating that one thing (the sea) is like another (dark wine), the metaphorist reconstructs the associative programming that automatically goes on in our brain when we experience that thing (the stormy late-day Mediterranean).

Epstein explores metaphor's "role in reifying the mind's associative network" in Proust's *À la recherché du temps perdu*. Proust's aim, Epstein asserts, is to make palpably explicit metaphoric relationships that are usually experienced only implicitly, to render evocative what we are only opaquely, tangentially aware of. "The tissue of metaphors that he creates acts as a symbol for something that cannot itself be directly represented—the network of associations, expectations, and understandings that usually express themselves only as an emotional overtone in the fringe of consciousness. In this way, Proust aimed to capture the essence of things—a stable structure of knowledge and memories that he believed existed 'outside of time.'"

The critical function of the novel, then, is not to convey the emotions the author felt, but to convey the interaction between the author's mind and the specific phantasmagoria of remembered events that brought forth these emotions. Epstein proposes three neural/cognitive mechanisms. First is a global process by which information from different cortical regions can be combined to form a momentary stable and coherent "nucleus" that is the focus of each single thought. Second is a network of associations supported largely by the medial temporal lobe's memory storage system that determines the relationship between the current nucleus and other potential thoughts. And these associations, finally, are shaped by executive control operations in the prefrontal cortex.

In Epstein's model, the hippocampus provides the "roads along which the stream of thought" will course. But there must be a "driver" that chooses the particular path to take: this is the dorsolateral and anterior PFC. Another PFC area then evaluates the desirability of following one stream or sequence of actions and not another: the orbitofrontal cortex, along with input from perhaps the amygdala. The PFC as a whole system may monitor the imagined sequence of events for narrative consistency: "given what we know about how the world works, could this episode have happened this way?" A special version of this, says Epstein, may monitor the imaginary sequence for autobiographical consistency: "given what I know about my past, is that event likely to have occurred?"

If the PFC "judges" the parts of a work of art as "fitting together," we (either artist or observer) experience a rush of satisfaction similar to the one we get when a new thought fits with previous thoughts and trains of thoughts projecting into the future. Epstein is influenced by the cognitive scientist Bruce Mangan, who in a 2001 paper, "Sensation's Ghost: The Non-Sensory 'Fringe' of Consciousness," described this sensation as a feeling of "meaningfulness" experienced at the far edges of awareness.[19] Taking pleasure in something "beautiful," we attribute this meaningfulness to its "surface features"—a breathtaking view from the summit, a searing line from a poem, mind-blowing colors in a painting—but, in fact, Epstein and Mangan claim, it arises from our nonconscious judgment of "a good fit."

Epstein calls upon Zeki's quasi-Platonic idea of the constancies in visual experience—the permanent, essential characteristics of objects that visual art taps into—and then proposes that this network-of-associations model applies to narrative as well. We experience the link between the story's described sensory surface and underlying associations that cannot be directly experienced or articulated. We say the power and excitement of a story derives from these sensory surfaces, (for instance, the literary "style" of the piece). But, says Epstein, it actually comes from "its reference to the unconscious associative network beneath." Whereas in visual arts the referenced networks may represent objects, scenes, colors, and shapes and the relationship between them, in narrative art these referenced networks primarily represent episodes, events, motives, and character and the relationship between them. In dramatic performing arts—music, movies, theater, dance—both operate simultaneously. Think of the multitextured reverberations of Kubrick's films, or even in *The Sopranos*, a TV series that operates on several levels at once.

"At any one moment, we experience the conscious correlate of a particular neuronal firing, but not the networks of neuron connections that controls the transition from one firing pattern to another," says Epstein. Art's goal is to "indirectly represent these neuronal networks, which control the stream of thought but cannot themselves be directly experienced. Essentially art is a trick that allows us to indirectly convey the structure of our minds." A successful work of art is an especially delightful mirror of sensory, emotional, and cognitive neural functions.

6

SILICON MINDS
The Rise of the Machine Genius

Machines endowed with a certain intelligence are becoming increasingly commonplace in the human landscape. In 2004, twenty-one thousand service robots were milking cows, handling toxic waste, and assisting in surgical procedures. Detailed GPS tracking programs advise motorists on the best route to their destination. The population of domestic robots is expected to surge sevenfold by 2007, as people buy machines to mow lawns, vacuum floors, wash windows, clean pools, and do other housework. By 2010, the army of cyborgs may reach a million strong.[1]

Electronics inventor and AI guru Ray Kurzweil, among others, has long predicted a human-computer mind meld. Yet even as we more tightly embrace our artificial agents, we are disquieted about what smart machines imply about our own authority. The movie *I, Robot*, like other science-fiction narratives that affirm human hegemony over a machine that develops a "will of its own," reflects a deep ambivalence about AI. At root, perhaps, is the unease that as machines grow more sentient, we will discover the robotic aspects within ourselves, offers Duke English professor Priscilla Wald. An expert on science in popular culture, Wald claims that we're threatened by how relatively simple it is to program a computer to behave with the "uniqueness" of a human; relatively simple because our behavior does in fact follow predictable patterns. The more artificial intelligence matures, and the more humanlike our machines become, the more we comprehend how machinelike we are. "Many people," says Wald, "find that deeply disturbing."[2] But not everyone.

As William Burroughs said presciently in *Naked Lunch*, the study of thinking machines may teach us more about the brain than we can learn from "introspective methods." Creating artificial intelligence de novo holds the powerful allure that in striving to build silicon replicants we might penetrate the great enigmas of life. Some brain scientists are dissatisfied with the slow progress of "introspective" methods, such as cognitive research, in yielding insights into large-scale mental phenomena such as the mind/brain problem, consciousness, learning, or how even smaller circuits actually work. "My frustration with psychological constructions is that the definitions feel mushy, ethereal, evanescent. It's difficult to put your finger on many of them and get anything precise," complains Todd Braver. "How do you consider a construct described in psychological language and translate it into a physiological system? How can you get neurons and axons and chemicals that flow between each other to realize what we call, say, 'attention'?

"There's a kinda dirty secret in cognitive psychology," he goes on: "when you tell people to do a completely esoteric, crazy, and arbitrary task in the lab, and you give them a minute of practice, they do it easily. And we don't know what happens! Clearly these instructions are completely novel, and the mind probably doesn't have representations for them. But we do have this PFC that, by activating the appropriate combinations of information, can learn something new. How can we do this? Our computational models," he adds, "are confirming a method."

On a hot September afternoon in his Princeton office, Jon Cohen, too, is talking about machine intelligence. "This is one of my background fantasies," he admits. "In the movie *2010*, you meet the guy who built Hal. Well, there's some part of me that would love— actually, there are many people in computer science who would love—to take a powerful machine and make it operate as a person does. It would be especially wonderful if you've built up this machine by using theories of how people function, as opposed to trying to invent it whole cloth.

"Of course there may be many ways to get a machine to imitate people without knowing how people do it," he adds. "But if you base your ideas on data about human behavior, how the brain is organized and functions, then replicate it in a device that does what people and brains do, the device becomes a more persuasive argument that you've got a good model. Creating intelligent life may be just the ultimate narcissistic fantasy. I don't necessarily imagine that I will be able to do it, or even be part of the immediate effort to succeed. But that it is possible excites me; and that some effort someday will succeed, in part based on the work we've done, is a rewarding aspect of our research. Yeah, it would be tremendously exciting."

The connectionist's dream is a flexible thinking machine, developed on a parallel distributed processing platform, engineered from blueprints based on prefrontal networks. In reality, the elements are fairly simple: arrays of processor "units" that simulate the synaptic alliances of neurons. Once running on their own, without human intervention, training indefatigably, colonies of units hook up, acquire connections. The system "teaches itself" to "think about" its performance in terms of the task's most outstanding features and goals; it associates linkages; it learns; it adapts. It drives itself inexorably forward, seeking internalized silicon "rewards." The desire to be rewarded may be the machine's gift from its carbon-based creators. Yet goaded by this central motivation, this "desire," if you accept Kent Berridge's concept, the neural net will embody an unslakable machine lust. The little units within the box will become as greedy as any driven living creature. Not "wanting" to be deprived of reward, the machine will work more indefatigably than any blood-and-tissue organism to do what's necessary to get that prize.

The model's human creators, meanwhile, observe this struggle from their paper-strewn desks. If they want to change the machine's "conduct," make it crazier, they tweak the processing units, strengthening one pathway or weakening another, then chronicle the results. With a Muybridge-like obsessiveness, connectionists run thousands of trials, comparing minute changes in each iteration, fluctuations in electronic units against the wildly less controllable behaviors of organic, wetware networks.

In contrast to the infrastructure undergirding animal thought, computational models today seem facile, toylike. Running a neural net may resemble playing with a dollhouse version of the brain. But behind this gloss of simplicity, dense layers of theory and empirical knowledge inform each model. If science possessed a complete wiring diagram of the brain with a definitive characterization of every brain-specific gene and protein, even this information would not suffice to reveal how the brain works. For if it were possible to assemble a model brain, faithful to the real thing, one real neuron, glia, and astrocyte at a time, one would enter a labyrinthine funhouse such as Borges's "Library of Babel," containing a near infinity of all possible synaptic connections, in which no two are identical. It would not be a simulation or translation or virtual brain. If against all odds it worked, it would be a brain.

Right now, there are many varieties of detailed neural-net models, and their verisimilitude in simming specific, tightly controlled data sets is high. Yet even as they simulate the nitty-gritty of neurochemistry, record intracellular calcium levels and postsynaptic potentials, these models mainly serve to yield a narrow-focused standard science. In establishing a centrally operating "mind" in their simulacra, modelers like Randy O'Reilly, however, opt for a more aesthetic approach, creating a "bit of an impressionistic painting," according to O'Reilly, who also creates dreamy quasipointillist works of computer art. "We find those points of contact in the neural mechanism that are doing most of the work. We capture these in some simplified way, and create out of them, as opposed to just drilling down and overloading the model with all the detail we can find."

To bring coherence to the data Tower of Babel that is contemporary neuroscience, Cohen and others developed a kind of big-picture model, a loosely knit integrative field theory of PFC function that can be tested in neural-net models. In a 2001 paper with Earl Miller, Cohen developed a "guided activation hypothesis," or more simply, the top-down bias.[3] Upon the central principles of this schema connectionists are building increasingly elaborate neural-net models.

What is bias in the brain? The term conjures up the "bias" in media that bends the news to one political persuasion or another. But more basically, Cohen offers a simple metaphor: water flowing downhill into two different catchbasins; water coursing left or right fills either of two discrete pools. If you want to fill only the pool on the right, you erect a barrier that diverts the flow there. PFC biasing—sending the flow of activity to the place you want it to go—is central to the neural model of guided activity. "To rapidly learn a complex behavior," says Earl Miller, "you don't wire up your whole brain. You wire up this little representation of it in your prefrontal cortex. Once you are about to engage in your task, the PFC calls up this model, then sustains it in patterns of neural activity. Biases." These bias signals in turn feed back to the posterior cortex, which activates more pathways needed to carry out the task at hand. This core PFC representation—this proxy, mental movie clip, simulacrum, "stand-in"—for an idea, goal, or intention to act is the essential genius of intelligence.

The big-brain PFC can potentially run limitless numbers of these representations, each having different biasing influences on other brain systems, each producing a different patterned response, in various combinations. But not that many at once. The cost of bias in this multiplex PFC is a great jostling competition among the different microcosms—and potential for interference and confusion. One way the brain organizes competing scenarios is to "downsource" the less complex ones to lower-level sensory and motor systems. The PFC doesn't micromanage; it delegates learned behaviors (how to ride a bike) to faster, less attention- and energy-draining sectors in other parts of the brain. These learned behaviors—rigid, stereotypical, not readily adaptable to new situations—rely on "bottom-up" processing overseen by simpler stimulus-response networks. But life's exigencies demand a flexible mind, that a person not be stunned into inaction by the unanticipated. The PFC reacts quickly to unexpected and ever-evolving scenarios. Working memory, selective focus, multitasking, all depend for guidance on top-down biased representations encoded in patterns of activity in the PFC that, in turn, reconfigure other brain pathways.

To be flexible and self-organizing, the "mind" of a machine, too, must wield this prefrontal biasing capability. That is, the machine's "PFC" networks must be able to handle competing representations that come in several forms. One is ambiguity, uncertainty—in which the machine has several equally viable options from which to choose. Computationally speaking, this occurs when information streams "activate more than one input representation." Or the machine is faced with competition that arises when a number of responses exist, but a weaker right response must compete with a stronger, but wrong, alternative. Take the Stroop test, for instance.

Miller and Cohen offer another "Stroop-like" scenario to explain this competition. Imagine yourself pausing at an urban intersection. Your unconscious inclination is to look left before walking across the street, and this is usually the appropriate thing to do—unless you are in England. To avoid being blindsided on the streets of London, you must remind yourself to look right before stepping off the curb. Most tourists require a few taps from the prefrontal billy stick to suppress the hardwired act of glancing leftward. The PFC imposes biasing to redirect your attention to look right, while simultaneously inhibiting the stronger but wrong impulse to look left. And when you commit to looking right, and do it repeatedly during your walking tour, you reinforce neuronal firing, strengthen the corresponding wiring patterns, augmenting the newly biased connections between the PFC and other brain system activity that supports looking right in England. Eventually the representation is relegated to lower brain systems, so that you unconsciously look right before crossing the street. A robot based on a guided-activation neural net, too, could eventually "learn" to avoid getting pancaked at Piccadilly Circus.

Cohen's team, as we discussed, programmed top-down biasing into a machine that learns to perform the Stroop test. The neural net was rewarded for "choosing" color-naming over word-naming, even though its "natural" bias was to choose the word's name. So the machine's "color control units" eventually became biased in favor of the color-naming pathway. That is, after many trials the machine "learned" to suppress the stronger word-naming "impulse." A cornerstone of the Stroop neural net is what the modelers called "active maintenance in the service of control."

For a mental representation to have biasing clout, it must be held online as long as the machine is performing the operation. The Stroop machine's "working memory," then, is a continuum of "representations" defined by the relative strength of the pathways supporting "color-naming" as opposed to those carrying the competing "word-naming" information. To hold on to the weaker color-naming representation, the machine must keep its "PFC" top-down online "control units" sovereign in the face of the challenge from word-naming units.

To describe the PFC's control operations, Cohen and Miller offer another transportation metaphor: the PFC units act as switch operators in a system of railroad tracks—neural pathways—hooking up various origins (ideas, plans, etc.) to various destinations (responses, actions, etc.). The PFC's job is to keep the trains (activity patterns that haul information) chugging to their rightful destinations and to avoid collisions. If the track is clear, trains can zip from origin to destination without risk of crashing into other trains, the behavior can be carried out automatically and will not need prefrontal intervention. If two or more trains must cross the same section of track, however, some switching coordination in the PFC is required to guide them safely.

"We think of the PFC as an operator, with a little map of this railroad system that tells it which tracks must be open," says Miller. "The PFC holds up this model railroad line for the rest of the cerebral cortex to see, allowing it to send trains along the tracks accordingly. That activity pattern sets up top-down signals that bias what's going on in the rest of the system. This is a powerful idea," he avows, "because, if behavior is dictated by whichever activity pattern is maintained in the PFC, changing behavior is as easy as changing that pattern." The simplicity of this idea makes the possibility of flexibly sentient machines more realizable—even as it implies a more plausibly "machinelike" precision of human wetware mental operations.

"We liked the railroad track switching analogy, because it implies the PFC is not pulling the levers, but just making sure trains go in the right direction," Cohen adds, "perhaps switching tracks now and then to prevent one 'train of thought' from bumping into another coming from a different direction. That said, you can imagine that for this very complicated system of tracks, at any one point in time there are many different switches being switched. So multiple representations are perpetually in play, with the PFC switching this one to the right, that to the left, to accomplish the net effect of having all trains reaching their proper destinations." Mental disorders might be compared to trains that, lacking proper switching guidance, fail to reach their proper destination, whether due to derailment or collision.

The PFC Stroop-playing machine of the late 1990s had elementary switching systems. Cohen and company soon added "anterior cingulate" units, which served as a watchman, to call the PFC to vigilance during ambiguous and competitive switching operations. The ACC units, explains Cohen, say, "'Wake up, switchman, you've got to do your thing!'" Thus the conflict monitor, the ACC, added another layer of cognitive control and flexibility to the neural net. But how the ACC knew to alert the switchmen remained unresolved.

As a part of their ACC modeling, the scientists realized that attention and inhibition—the two pillars of cognitive control and intelligence—might be designed into the machine as opposite sides of the same coin. Attention would be the effect of throwing units' weight toward relevant information (focusing on color-naming in the Stroop). Inhibition would cut off the competing signal at the knees (suppressing word-naming). The neural-net PFC units, with help from ACC units, would bias representations, would switch on and off the desired combinations of attention and inhibition trackways, and the machine would learn to make the right choices on time, according to plan.

But a conceptual elephant still occupied the room: how does the PFC "know" what representations are to be held online, and how long they are to be held? It's the problem of infinite regress again, the Wizard of Oz sitting in the brain's balcony whispering orders through a wire. No, this homunculus had to be exorcized. "Dopamine is our way of banishing the homunculus problem," Earl Miller declars. "It is a perfect teaching incentive." If in living systems, dopamine is inextricably tied to reward, might neural nets, too, be programmed to "crave" simulated dopamine rewards just as deeply in their silicon units?

The subject of prodigious biological scrutiny, dopamine was nailed the "reward signal" when, in the 1990s, investigators such as Wolfram Schultz, now at Cambridge University, showed that dopamine neurons that project to the PFC from the lower reaches of the brain, such as the ventral tegmental area (VTA) and the substantia nigra, could behave as molecular teaching devices. Schultz and others demonstrated that changes in dopamine firing enhance

learning by reinforcing reward in animals, and the dopamine learning concept could be applied to computerized neural nets. (On
Schultz's home page is posted an array of sushi platters with a sign
saying "reward" in big letters.)

Some of Schultz's research relating to computerized neural nets
is built on earlier reward work. In 1972, the animal learning theorists Robert Rescorla and Allen Wagner published a seminal paper
presenting a mathematical model for aspects of classical conditioning. Theirs was a computer program based on Pavlovian stimulus-
response behavior, where an animal learns to associate a tone, flash
of light, or other cue with a food, juice, or other reward. The
Rescorla-Wagner work was a big deal, encapsulating whole bushels
of learning phenomena in a simple set of equations.

By the 1980s, most neural-net learning algorithms had become
extensions of the Rescorla-Wagner learning rule. Since then, neo-
phyte modelers everywhere design their rudimentary classroom
sims using the Rescorla-Wagner learning rule. For his PhD thesis
back in the 1990s, Todd Braver, for example, trained a computer
model that was initially "ignorant" about the relationship between
stimulus and reward to eventually "understand," like Pavlov's dog,
that a particular cue meant future reward. In the late 1990s, Schultz
proposed that modelers could use simulated squirts of dopamine
signals to drive the Rescorla-Wagner learning rule in computer
neural nets.

Around the same time, neural-net modelers were attempting to
add anticipation of future reward onto the machine narrative.
Again dopamine served as the signal to do this. When you present
a living creature with a reward, dopamine neurons fire and inform
the cortex about the nature of this reward. Then if you repeatedly
present the animal with an arbitrary cue, a ringing bell, say, just
prior to a reward, the dopamine neurons eventually cease to fire for
the reward, and begin to fire for the ringing bell. Thus the animal
learns that the ringing bell predicts a future reward. One can "walk
back" the eventual reward to earlier and earlier cues, and the
dopamine neurons will fire for the earliest event that prefigures the
reward, the first cue in a cascade of cues.

Think of how you learn a task. You try out different strategies,
occasionally hitting upon the (rewarding) right method or correct
answer. "When that happens," says Miller, "dopamine gets squirted

up, from the VTA through the striatal-basal ganglia-thalamus network to the PFC, and causes connections to be formed among the PFC neurons. This causes the formation of little models in the prefrontal cortex—the patterns of the railroad tracks, if you will—that you need to solve the task. By trial and error you learn the basics of how to get the reward, then you add layers of complexity on top of that. This is exactly what these dopamine neurons seem to do: 'chain in' more and more complexity along a sequence of events that ultimately lead to reward."

Yeah," Cohen adds, "it feels good to know you are going to be rewarded."

Moving the squirt of dopamine back from the instant of getting the reward, to various expectancies of getting it, links the reward picture in your PFC to longer and longer durations, significant delays in time between the first expectation and, finally, the reward. Thus as dopamine neurons start to chain backward in time, associating more and more complex events, they paint richer and more detailed information into this reward picture. Cohen calls this dynamic chaining phenomenon "bootstrapping," because the neural-net system pulls itself along by its own bootstraps. No need for a homuncular puppetmaster yanking the strings. That dopamine is the linking mechanism of learning across time is a powerful idea—one can see entire human civilizations built upon it, adding layers upon layers of associated phenomena upon an original stimulus-reward. Think of our last nationwide Big Reward project, the space program: from the little rewards of small suborbital rocket launches to the Moon, Space Station, and perhaps, someday, as David Chappelle puts it, "Mars, bitches!"

"A lot of people criticize our approach, saying we're making dopamine into a homunculus," laughs Randy O'Reilly. "But actually, by itself dopamine isn't all that powerful. The key insight the computer models provide—and their overall intelligence—comes from the interaction of all the different parts. Sometimes we're not as careful as we should be in making that point clear. Dopamine becomes a key part of the model only when it becomes integrated with the top-down systems that drive dopamine signals in ways that make the system smarter. Dopamine signaling is the end result of a lot of neural computation, a convergence to drive the global signal."

Dopamine teaching signals can also switch off learning trackways that do not provide reward. Normally, dopamine cells hum along at a baseline of activity. It's the unexpected reward—or the sign predicting a reward to come—that evokes spikes of activity in dopamine cells and increases their rates of release. Take this quasi-moronic example of reward learning: you jiggle the rusty old vending machine knobs this way and that, and eventually, the package of M&M's descends, at which point dopamine squirts within your cortical learning areas. By enhancing the "go" firing-wiring patterns in this pathway, you encode your anticipatory strategy for how to get the candy the next time.

Should you jiggle the knobs the wrong way on the vending machine, you get no candy, and there is no spike of dopamine. During this "no-go" signaling, dopamine firing drops below the baseline hum. In a "no reward" situation, lack of a dopamine burst disinhibits the usual silent "no-go" neurons, allowing them to fire along other pathways. The "no-go" signal thus teaches you to reject this stratagem in the future. The same thing happens when, based on successful jiggling experience, you expect to get the M&M's, but they fail to drop down. You go through the "right" moves, but the candy remains in its niche. "No-go" signals fire vigorously, and thus instill in your PFC a revised mental model—to bring a baseball bat the next time.

Activating a PFC representation invokes a goal or rule that can be flexibly switched on or off as circumstances demand. People switch fairly effortlessly from one representation to the next, as new goals replace old. But eventually, Cohen adds, "the PFC is filled up, and so is our information-processing capacity." Since capacity to run simultaneous executive programs is limited, PFC networks have evolved gating devices. There is continuous gating tension in the PFC: a system that keeps stuff in, keeps stuff out.

Jon Stewart once asked in the context of the Watergate scandal, "Who hands out the 'gates'? Is there a 'gate' gate?" Machine modelers, too, were faced with the question: who manages the gate, if there is no homuncular gatekeeper? "The gating system sustains in your PFC a current representation—of naming color over word—allowing you to execute this Stroop task without distraction," says Cohen. "You ignore your hunger pangs, the sounds outside the room, the fact that you'd rather be somewhere else doing something else. But you don't spend the rest of your life naming colors every time you see color words. At the appropriate moment, you let go of that plan's representation, and let something else take its place—getting paid, going home, finding something to eat."

Tension within the gating system presents design problems in self-organizing machines. "If you make this PFC system highly resistant to distractions," says Cohen, "you run the risk that it will just keep doing this thing forever. On the other hand, if you make it too labile, then it won't be able to hold on to the representation for the duration you want it to. So there's this delicate balance between keeping going against competing alternatives for the right amount of time, and then, miraculously, at the right moment suddenly being flexible enough to snap out of it and be attentive to something else."

A machine designer can simulate gating by logical programming. "You can train a network to sustain the color-naming representation against interference," he continues. "You can train the model on all kinds of alternatives, render all the circumstances in which they might arise. But it's computationally weak; it's a very brittle method. You wouldn't design a program or robot that way because it takes too much computing. It can be done, but"—he utters the damning words—"it's not robust." So again dopamine came in handy to drive the self-organizing gating mechanisms.

At first the dopamine story was "rather simple-minded" and didn't take gating into account, Cohen admitted. "You replace plan A with plan B, via a dopamine gating system that controls updating and maintenance of PFC internal representations. But goals have subgoals. I have to get to work, but first I have to put on my jacket, go out and start the car, and so on. Every time I go to gate one of these subgoals, I'm at risk for losing the higher-level plan I'm trying to maintain—getting to the office. Every time I update subgoals, what keeps competing possible alternatives from rushing in to interfere with the superordinate goal?"

The modelers realized they could use the "chaining backwards in time" feature of dopamine learning signals for gating the neural nets. They permitted the program to murmur to itself, "For the moment, we will not allow any information to hit these units; they will be preserved. But when we are finished with them, we'll open the gate and allow the strongest, loudest intruding alternative to take over." Cohen's group installed a gating application in the

Stroop model to keep color-naming as active within the PFC as long as necessary. When finished, the machine would open the gates to other representations of reward they had built into the machine—simulated hunger pangs, say, or intent to get paid.

But, again, how would the machine "know" when to open or close the gate, when to flip its "mental" representations? The dopamine units, the modelers decided, could serve as gatekeepers, as they learned through trial and error. Through trial and error, dopamine would steer the system toward figuring out what choice to make to get the reward, and then how to switch off the gate when the reward was delivered or the strategy no longer panned out. Since dopamine signals link action to reward, and are predictive of future reward, dopamine could signal the machine when to stop the behavior. "We showed that if you put those two together, learning and switching, working through the same mechanisms, suddenly you've got a system that can spontaneously self-organize," says Cohen. Maybe not a robot, but at least a device that can learn to execute short-term working-memory tasks such as the Stroop.

The neural net was still primitive, and floated in programming space unattached to "subcortical," "limbic," or "brain-stem" simulations, real neural systems that powerfully drive the living brain. The machine's dopamine did not "come" from anywhere. For the PFC machine to learn, to self-organize, via electronic activity patterns, the modelers had to endow it with a way to "know" what information is relevant and what is not, when to bring in new information, when to discard old, when to protect the PFC's internal picture of its goal from distractions, how to assign rightness or wrongness to events. So a next step was to design a virtual, self-organizing, PFC learning machine with inputs from simulated dopaminergic subcortical wellsprings that, in the real brain, work through complex trackways of the basal ganglia.

For over five years, Randy O'Reilly at the University of Colorado had been designing and refining a simulated basal ganglia neural net to hook up to the PFC machine. The subcortical basal ganglia (BG) have long been considered disinhibitors, brake releasers, of frontal lobe motor actions. They detect the right contexts for performing physical actions and give the "go" signal, enabling the frontal lobes to execute these actions at the right time. In machine architecture, O'Reilly extended the BG's domain to confined cognitive events such as working-memory simulations. It was to further explore the BG's role in complex large-scale cognitive phenomena that he built the BG neural net. Today his group's PFC–basal ganglia neural net, he says with no false modesty, "probably represents the high-water mark in complexity of models simulating complicated psychological tasks."

A founding member of the top-down club, O'Reilly is the go-to guy for simulating large-scale cognitive phenomena. "Of the three of us, he's the most skilled modeler," says Todd Braver. "He's amazing that way." Since reading about artificial intelligence in high school, O'Reilly wanted to build robots and sentient machines. "I had this little toy computer back then, and tried to make some kind of AI program for it," he recalls. Enrolling in Carnegie Mellon's legendary computer science program, he could have gone the traditional, program-dominant, symbolic artificial intelligence route. "Although those models weren't right in a lot of ways, it was easier to represent a lot of complicated knowledge in them, whereas with neural nets you have to train them. That training really adds up. I have networks that take days and weeks to learn!"

But O'Reilly, like Braver, wanted to design thinking machines that paralleled real brain systems, in part because conventional psychological theories fail so miserably in revealing how real brains think. He points to hypotheses based on "box and arrow" models with labels for processing operations ("object recognition," "short-term memory buffer," and "lexicon") connected by arrows that indicate "information flow." Other types of theories, he says, relied on the brain-as-computer metaphor, with the cortex having a "CPU" and various short- and longer-term storage buffers. Real brain networks, however, operate in a massively parallel fashion where multitudes of neurons work together, each making a small contribution. "Memory and processing are embedded inextricably within each of these neurons, instead of being segregated as in a standard computer. Information is not simply passed around but is embedded in distributed patterns of neural activation that directly influence processing in other parts of the brain. New kinds of thinking are required to understand how these neural systems behave."

With his partner the psychologist Yuko Munakata, O'Reilly has written a neural-net software package, delineated in their textbook

Computational Explorations in Cognitive Neuroscience: Understanding the Mind by Simulating the Brain. He named the software Leabra, a balance of computation and neuroscience, inspired by the astrological Libra, the balance scales. The neural units in Leabra simulations use equations based on ion channels in cell membranes that govern the behavior of real neurons, and Leabra-based neural networks incorporate anatomical and physiological properties of the neocortex.

The PFC machine was now ready to be hooked up to O'Reilly's virtual basal ganglia subcortex. Indeed, its dopaminergic gating mechanism now demanded feedback-feedforward circuitry from the subcortical cellar. Without benefit of some BG-like moderator, the PFC, swamped by competing scenarios, would simultaneously try to execute all of them. This would lead to "high amounts of motor or cognitive interference": the machine correlate to a human's physical or mental confusion. If the sim possessed basal ganglia units, reward responses in the virtual frontal lobes would load easier and be faster-firing, and unrewarding responses would be more quickly suppressed.

To make the basal ganglia units workable, O'Reilly divided this system, so multiplex in the organic brain, into two big units. He labeled one part the "actor" system, composed (in the living brain) of parts of the thalamus, the dorsal striatum, and globus pallidus that project to the PFC. The "actor" system, says O'Reilly, is one "that actually does stuff." The other BG section, which he called "the critic," roughly corresponds to BG subsystems that handle reward—such as the nucleus accumbens and the amygdala. "The critic trains the actor; it critiques what the actor does, then retrains it to do better next time." The PFC's top-down superstar director is influenced by the BG's actor, who is coached by the critic. "So it's all one big integrated circuit in our model, like a dog chasing its own tail, everybody influencing everybody else," says O'Reilly.

The machine's first task was to learn to "win" at a fairly complicated working-memory challenge. More arduous than the Stroop, the task involved two tiers of goals and resembled rounds of a simplified gin rummy or poker game. Called the "1, 2-AX task," it works like this: numbers and letters (1, 2, A, X, B, Y) stream in sequences, and the participant (human or silicon) must figure out how to "hit" one of two target combinations, either AX or BY.

Which one is the "winning" target is determined by whether a 1 or a 2 popped up most recently in the sequence of letters and digits. If the player last "sees" the number 1, he will eventually learn that the target is AX; if the player last "sees" the number 2, he'll subsequently deduce the target to hit is BY. Thus the number 1 or the number 2 must be held "in mind" over repeated trials, while the participant puzzles out the correct hit in the series of letters flowing by.[4]

O'Reilly calls this holding online of 1 or 2 the "outer loop" in the goal hierarchy. In the machine's units, this outer loop corresponds to anterior, frontopolar sectors of the human PFC—such as BA 10—that are involved in more abstract, "internal," metalevel processing. Intelligence, says Braver, is the ability to manage competing and hierarchically structured goals and those that require frontopolar, outer-loop processing units. These are reasoning and strategy-developing processes in which one needs to say, "If I did A, as opposed to B, then what will happen?" while you hold in mind A and B in relation to the long-term goal.

Watching for and finding the right target in the task—either AX or BY—in the stream of possible hits, however, requires faster, more dynamic thinking—the "inner loop." In the PFC-BG machine, this inner loop corresponds to the human's dorsolateral PFC zones where more nitty-gritty, subgoal, transient, dynamic, and external working-memory processing takes place. Here inputs from sensory systems need not generate a sustained mental image. Humans typically have all levels of this hierarchy operating and integrated simultaneously. The machine, too, must therefore rapidly and selectively update its inner-loop letter information, while sturdily holding online in its outer loop the representation about the 1 or 2 rule of the game.

The machine was initially totally ignorant of the task, the rules, how to score or get rewarded. Since the sequence of letters and numbers the model "saw" was unpredictable, the model had "no idea" what would be a rewarding hit. After it started playing the 1, 2-AX task for a while, it began to get rewards when it happened to "stumble across" the right answer. Through tirelessly repetitive trials, it learned to do the right thing—learned the predictive power of knowing the rules in its little electronic microcosm. The machine PFC units had to do two things: the inner loop pursued the correct target, while the outer loop "held in mind" what rule was valid at the moment. The basal ganglia units updated the rules by using dopamine-driven "go" or "no-go" signals to switch back and forth as the rule changed.

The first part of the machine's learning program—in which it was rewarded with a dopamine squirt when correct—is "nothing terribly new," says O'Reilly, "it's classic Pavlovian conditioning, the Rescorla-Wagner rule, just that." This corresponds to input from the living brain's nucleus accumbens reward "units." The "evolutionary addition" to the PFC-BG machine lay in its "perceived value system." This splitting of reward signals into actor and critic endowed the machine with a more flexible learning apparatus. The critic does not fire when the reward is delivered, but rather evaluates and predicts how likely the chances are the actor's actions will lead to another reward in the future. If the cue that previously delivered a reward delivers again, the critic takes note, shooting a dopamine "go" signal to the actor system on the next round.

If it determines there is no dopamine reward, the critic sends the actor the "no-go" signal. If the machine opts for a reward with the no longer relevant rule, it gets a metaphoric "punishment" by the critic, and learns not to make that mistake again. The difference between the critic's predictions and the actual outcome in reward (positive feedback) or "punishment" (negative feedback) in each round of the game is what drives the machine's learning. When the critic learns to predict reward or no-reward with perfect accuracy, the actor then switches from rule 1 to rule 2 to another with a seamless gating mechanism. And learning ceases.

Successfully run on a wide range of tasks, the PFC–basal ganglia model powerfully demonstrates flexible machine learning on complex working-memory tasks, and showcases the functional advantages of a gating program for working memory. This model, states O'Reilly like a proud father, is the first biologically based machine for controlling working memory that can match wits with and beat the learning prowess of more code-heavy and biologically implausible AI devices. Since the PFC-BG machine is based on biology, it can be harnessed to explore real mental disorders just by perturbing the machine's parameters. Its simulations enable it to test various theories of attention deficit hyperactivity disorder, for instance.

Recently Michael Frank, in O'Reilly's lab, used the neural net to

gain insight into the role of dopamine in Parkinson's disease suffer-
ers' cognitive problems. In these patients, the substantia nigra, a
subcomponent of the basil ganglia, degenerates and produces less
and less dopamine. O'Reilly and Frank studied patients who were
both on and off their regular dopamine medications and found
thinking behaviors that paralleled those of the neural-net model.
Clinicians have noted that patients on medication learned better
from positive feedback (rewards; "the carrot"), while those off medication actually learned better from negative feedback (mistakes;
"the stick"). No theory of Parkinson's disease had previously been
able to account for this odd phenomenon. "Our basal ganglia
model made this prediction about learning in Parkinson's that
nobody else was making," says O'Reilly. "And it turns out to be confirmed." Both carrot (dopamine reward) and stick (no dopamine)
feedback are central to instilling appropriate behaviors in healthy
people. When making a decision, we consider both the pros and
cons of various options, which are influenced by carrot-and-stick
outcomes of similar decisions made in the past.[5]

The PFC-BG machine serves as a bridge between powerful
abstract computational mechanisms and living brains. Previous AI
machines based on the wetware brain had proven woefully incapable of simulating problem-solving and reasoning operations that
require dynamic working memory and gating over a time scale
ranging from seconds to minutes. The PFC-BG machine can do
this. But it is just a start. A primary challenge to any connectionist
modeler, O'Reilly admits, is accounting for the extreme speed and
flexibility of human thinking abilities. Instead of needing hundreds
of hours of training trials to learn elementary tasks like the 1, 2-AX,
humans pick it up almost immediately based on verbal communications. Our model, he adds, may more closely simulate the working-
memory performance of "a relatively gifted and motivated
monkey."

To create silicon sims of other aspects of the human PFC, Todd
Braver and other members of the modeling group designed some
variations on the PFC units, such as a virtual emotion reactor.
This machine begins to simulate the divisions of cognitive-
emotional labor. In one elaboration of the model, the PFC sim is
subdivided into ventrolateral (orbitofrontal) and dorsolateral sectors. The orbitofrontal cortex may be "responsible" for "hotter"

social, emotional pathway switching, whereas the dorsolateral PFC might handle "cooler" more cognitive control tasks. Because hotter information is likely to elicit reflexive and inappropriate reactions, the ventral OFC units in an emoting machine might play more inhibitory switching roles, serving to bias cooler but relevant processes against hotter, stronger competing alternatives that often prove less rewarding in the long run. The cognitive processes that engage more dorsal regions are less likely to engage "asymmetries of strength," as Jon Cohen puts it, and so their competition is likely to be "less fierce."

O'Reilly, like Braver, sees this PFC machine divided less by the biasing drives of emotion than by hierarchies of abstraction. "Our levels of abstraction idea is a nice, simple theory. Ventrolateral parts encode the most detailed stimulus-specific representation," he says, "corresponding to our 'inner loop' in the PFC-BG model. Dorsolateral sectors manage slightly more abstract representations engaged with categories of plans, events, and actions. Another take on it is that the machine's ventral levels are involved in emotion and affective processing whereas dorsal units are more sensory and cognitive. The emotional area tends to be more stimulus-specific: more emotional representations are associated with the specific color red or blue than with the abstract notion of color. That's one way of reconciling these two views."

The artificial PFC's hierarchies of abstraction would determine which of its units would be switched on or off at any moment. Lower echelons, such as the orbitofrontal and the ventromedial, would tackle information processing for shorter time frames, and handle more emotionally charged data, while the dorsolateral and anterior PFC areas would manipulate increasingly long-term and internal representations. When modelers further differentiate the PFC units to coordinate the hotter-details/cooler-abstract coding with the inner and outer loops and the gatekeeping basal ganglia machine, this synchronized device might be deemed an "emotionally functional" neural net that learns to curb its enthusiasm for an "impulsive strategy" that yields a minor short-term reward, and opt instead for a more restrained and "considered plan" that yields a bigger prize over time.

But an artificial brain that learns, plans ahead, and reacts flexibly would also need long-term "prospective memory" units of the "future," as well as memory storage and retrieval systems: it would need a hippocampus. A hippocampal subsystem, says Jon Cohen, provides the context for picking up a plan "off the shelf" at the right time and place. Cohen provides one of his patented day-in-the-life scenarios: "I get up this morning, go down to the fridge, and . . . There's no orange juice! I'm already late to work, so I'll just have to remember to buy some later. My PFC creates a little representation of going to the grocery after work and stores it away.

"When the sun is setting, and the workday is over," he continues, "it is the hippocampus that binds together the plan stored in memory with the context and the timing for retrieving it." Coding within the hippocampus performs a kind of gluing function, creating an association between circumstances now and the PFC representation needed to respond appropriately. "You are leaving the office, and as you're putting on your coat, bingo! The hippocampus has encoded an association between this constellation of environmental cues, this context, and this internal representation about buying orange juice that you want the PFC to execute. Now you've got to hold the entire representation in the little microprocessors in your PFC.

"Now, do I keep that representation active all day? No way!" Chaos would ensue. Imagine how many other things the PFC would also have to hold online—infinite numbers of representations, just to get through the day. The PFC system is hardly infinite in capacity, so you have to let go of the orange juice representation. But how do you get it back at the appropriate moment? "I'm putting on my jacket, and I remember I need to get orange juice. I also remember I must make a left at the traffic light to go to the grocery instead of the usual right at the light to go home. On the other hand, I'm sure you've had the experience of making a new plan, then later doing the habitual thing instead, and saying, 'Oh shit, I forgot to do . . .' It's because you've failed to activate the plan."

Cohen, O'Reilly, and company think that by yoking the PFC–basal ganglia units and hippocampus they might create a self-organizing system in which context (workday ending) can, in the hippocampus's episodic memory, activate the stored representation (buy juice), and send it up to the PFC command (go to store). "The model shows how these two systems work nicely together: the environmental context elicits an internal representation that then leads to performing the appropriate new behaviors, as opposed to just responding habitually. I mean, the algorithms work," says Cohen. Imagine a houseworker robot that orchestrates and prioritizes several chores at once, and reminds itself to roll out to the store for orange juice before it switches itself off for the night.

O'Reilly has worked for years on a model of the hippocampus that will someday work in concert with basal ganglia and PFC units. Many in the field, however, shake their heads. They think the hippocampus and the BG present two competing learning systems; that the hippocampus specializes in spatial learning and episodic memory, while the BG deals with learning habits and physical skills. Once again, O'Reilly alters and expands this idea. His proposed model allows the parts of the basal ganglia and hippocampus to control different jobs. While the hippocampus binds together information and context into a representation (sun setting; time to get orange juice), the dorsal basal ganglia gates in this new representation (put on jacket; reroute driving map to include grocery store) and switches off the old representation (habitual route home).

In this proposed machine, the hippocampus links specific ideas, objects, and events with executive plans. And pathways thus activated are not rigidly tied to a specific task but can support a multitude of scenarios, strategies, goals, and so on. The hippocampus contributes to flexible behavior by presenting an archive of items that can be joined together ad hoc. Imagine you might need to park your car in a different place every day. The hippocampus can complete the representation by means of its context, and you remember the display of roses in the florist shop in front of today's parking space. The hippocampus prevents this specific representation of car-rose-shop from being flooded out by generalized knowledge about parallel parking that would not help you find your car on this very afternoon.

O'Reilly has a good working model of a hippocampus neural net; he's tested and validated it on tasks that challenge its specific functions. What his team hasn't yet done is hook up the silicon hippocampus to the prefrontal–basal ganglia machine. "We've talked about it for the last ten years," O'Reilly laughs, "but somehow it always ends up that there's more work to do on the pieces before it's time to put them together." Each piece is complete in its own right, but wired together the entity becomes immensely more complex. "It's one of those pragmatic challenges that has yet to be managed.

But just today I had a conversation with one of Todd Braver's post-docs, Jeremy Reynolds. He's again proposing to hook up these guys. So maybe he'll be the one . . ."

Attaching hippocampus to basal ganglia and topping it off with a PFC executive—that will be one thinking fool! "Exactly!" he replies, quickly adding that there are many other dimensions to a self-organizing silicon system: how it might simulate neurotransmitters beyond dopamine, for instance. Jon Cohen is beginning to model norepinephine systems, which in an AI machine might help balance focused attention, execution, concentration against wide exploration and a more distributed alertness. O'Reilly's group has also been thinking about how to "take on serotonin," the neurotransmitter so tied to mood regulation and to shifting between short- and long-term prediction of rewards.

Is there a kind of Moore's law for neural nets, whereby new machines evolve and replace old at predictable intervals? "If there is, it's rather slower than Moore's law," O'Reilly offers. "We've been developing the whole PFC-BG-hippocampus story for five years. The pieces are pretty well in place; there's increasing complexity in the way we are building up these models. I've been working toward building a more integrated cortical mini-brain that combines both PFC and hippocampus and a reasonable model of sensory and motor posterior cortical areas. We expect this to happen in somewhere between five and ten years."

Todd Braver also foresees an amped-up PFC machine within a decade. Braver is attracted to the "have your cake and eat it too" aspect of neural-net modeling. You can start out with a set of conditions for the prefrontal cortex to confront, modify one, and watch what the whole does differently than it did before. Tweak another variable and observe the result. "The way we do science in our heads is impoverished. Here we can put all the elements into a system, watch how it unfolds, and analyze it at different stages. We can add emotional inputs or make virtual lesions, damage different parts, and see how that affects the net. There's some empirical kick to it." The machine is an adult toy, yet both a partner to and a mirror of our minding selves. "It really is the closest thing to playing evolution," Braver admits.

Neural-net machines' growing self-organizing capability and autonomy increasingly demonstrate how feedback-feedforward systems can be functionally flexible and can teach themselves, can "bootstrap" independent of the programmer's heavy hand. As Jon Cohen remarked about his machine's ACC conflict-monitoring units, "the beauty of this idea is that it really requires no homunculus to get what appears to be a homunculus-like effect." The PFC–basal ganglia machine acts of its own "volition" in tiny increments within the confines of its silicon world. Similarly within the organic brain, the basal ganglia–thalamic loops provide a big-systems network for selectively updating the mental movies of our hopes and schemes that operate on PFC power, using neurosignaling agents such as dopamine to act selectively on recurrent pathways running from the PFC throughout the rest of the brain and back again to the PFC. Although the brain bases of systems of mind remain dazzlingly elusive, they are ultimately knowable. As we learn how these systems function as partially discrete entities and interface with one another in significantly greater detail, we shall also create ever-cleverer forms of sentient beings.[6]

Notes

Author's Communications

I talked to these people in the course of writing this book.

Steven Anderson—March 3, 2004
Mario Beauregard—December 3, 2002; June 14, 2004
Kent Berridge—April 27, 2004
Hilary Blumberg—September 9, 2004
Todd Braver—October 1, 2002
Montgomery Brower—December 26, 2002; January 8, 2003
Camilo José Cela-Conde—April 1, 2006
Kalina Christoff—October 30, 2002; April 2006
 Jonathan Cohen—August 15, 2002; September 4, 2002
Andrew Conway—March 4, 2004
Adele Diamond—July 10, 2002
Russell Epstein—February, 2005
Joaquin Fuster—April 2002; April 4, 2005
John Gabrieli—July 12, 2002
Vinod Goel—May 22, 2002; August 12, 2003
Stephen Goldinger—February 26, 2004
Patricia Goldman-Rakic—1986; June 2002 (interview with Douglas Stein—
    November 2–3, 1994)
Jeremy Gray—April 7, 2004
Edward G. Jones—February 7, 2003
Birgit Bork Mathiesen—March 1, 2004
Earl Miller—October 31, 2002
Brenda Milner—May 9, 2003
Randall O'Reilly—October 20, 2004
Adrian Owen—September 25, 2002
Michael Petrides—August 9, 2000; November 7, 2005
Bruce Price—December 13, 2002
Carol Seger—July 17, 2006
Arthur Shimamura—September 21, 2004
David Zald—February 10, 2004

269

Introduction

1. Burgess, P. W. (2000). Strategy application disorder: The role of the frontal lobes in human multitasking. *Psychological Research*, *63*(3–4), 279–288.
2. Jacobsen, C. F. (1935). Functions of frontal association area in primates. *Archives of Neurology and Psychiatry*, 33, 886. In Jack D. Pressman, ed. (2002). *Last resort: Psychosurgery and the limits of medicine*. Cambridge, England: Cambridge University Press, 37.
3. Fulton, John F. (1949). *Functional localization in the frontal lobes and cerebellum*. Oxford: Clarendon Press, 63.

1. Memory: The DNA of Consciousness

1. Cache memory itself is a metaphor. The pre-electronic use of the word "cache" refers to small stocks of food, ammunition, and clothes explorers and hunters hide in the field to have when they need them. This is the same in computers; one caches information to have it when you need it. In general, cache memory means faster than normal. It also implies more expensive than normal. A program might copy some information found on disk with access time in the milliseconds to a RAM memory location with microsecond access. The program will run faster accessing the cached information than if it has to fetch it from disk. For the same reason, hardware manufacturers will cache memory on the processor chip— nanosecond access—instead of relying on RAM with microsecond access. The information is where it's needed when it's needed.
2. Fuster, J. M., & Anderson, G. E. (1971). Neuron activity related to short-term memory. *Science*, *173*(997), 652–654.
3. Fuster, Joaquin M. (2003). *Cortex and mind: Unifying cognition*. Oxford: Oxford University Press.
4. Rao, S. C., Rainer, G., & Miller, E. K. (1997). Integration of what and where in the primate prefrontal cortex. *Science*, *276*(5313), 821–824.
5. Freedman, D. J., Riesenhuber, M., Poggio, T., & Miller, E. K. (2001). Categorical representation of visual stimuli in the primate prefrontal cortex. *Science*, *291*(5502), 312–316.
6. Nieder A., Freedman, D. J., & Miller, E. K. (2002). Representation of the quantity of visual items in the primate prefrontal cortex. *Science*, *297*(5587), 1708–1711.
7. Ingram, Jay. (1997, May 16). Jay's brain. www.exn.ca/main/jaysbrain/jb-19970516.cfm.
8. J. R. Stroop seems an unlikely individual to have made such an enduring contribution to the science of cognition. Born in rural Tennessee, he grew potatoes to finance his college education in Nashville. (See Colin MacLeod, psychologist at the University of Waterloo, for the definitive account of the Stroop effect.) Stroop was teaching psychology at Peabody University when he created the test, and there he remained for most of his career. The Stroop test became immediately popular. For "Brother Stroop," though, Christianity had a far greater pull than science. From his

youth onward he wandered as an itinerant gospel preacher, accepting payment for sermons with a chicken or bag of potatoes if at all. Stroop continued to teach psychology, but he never followed up on his early research on the "effect." Although he was aware, thirty years later, of the test's impact, he claimed to have no interest in the ongoing studies launched by its robust phenomenon. Stroop died in 1973 at age seventy-six. In 2002 scientists descended on Vanderbilt University, of which Peabody is a unit, to partake in a celebratory Stroopfest.

9. For more on the consciousness question, see Wegner, Daniel. (2002). *The illusion of conscious will.* Cambridge, MA: MIT Press; and Searle, John R. (2004). *Mind: A brief introduction.* Oxford: Oxford University Press.

2. Reason: Logic, Laughter, and Looking Within

1. Langewiesche, William. (2002, September). American ground: Unbuilding the World Trade Center, part two: The rush to recover. *The Atlantic Monthly, 67.*

2. Goel, V., & Grafman J. The role of the right prefrontal cortex in ill-structured planning. *Cognitive Neuropsychology, 17*(5), 415–436.

3. Legend has it the Tower of Hanoi puzzle, a.k.a. the Tower of Brahma, or the End of the World, was invented in 1883 by the French mathematician Edouard Lucas. He was inspired by a tale of a Hindu temple where at the beginning of time, priests were given a stack of sixty-four gold disks and assigned to transfer them from one of the three poles to another, with the same rule—a large disk could never be placed on top of a smaller one. When they finished, the temple would crumble into dust and the world would vanish. In fact, the number of disk transfers necessary is $2^{64}-1$, or 18,446,744,073,709,551,615 moves. If the priests worked constantly, making one move every second, it would take slightly more than 580 billion years to accomplish the job. Hofstadter, R. Douglas. (1983). Metamagical themas. *Scientific American, 248*(2), 1622.

4. The Schiavo is observation thanks to Ed Kilgore, TalkingPointsMemo.com, March 20, 2005.

5. Goel, V., Gold, B., Kapur, S., & Houle, S. (1998). Neuroanatomical correlates of human reasoning. *Journal of Cognitive Neuroscience, 10*(3), 293–302. Goel, V., Buchel, C., Frith, C. D., & Dolan, R. J. (2000). Dissociation of mechanisms underlying syllogistic reasoning. *NeuroImage, 12,* 504–514.

6. Noveck, I. A., Goel, V., & Smith, K. W. (2004). The neural basis of conditional reasoning with arbitrary content. *Cortex, 40* (4–5), 613–622.

7. Goel, V., & Dolan, R. J. (2003). Explaining modulation of reasoning by belief. *Cognition, 87*(1), B11–22. Goel, V., & Dolan, R. J. (2003). Reciprocal neural response within lateral and ventral medial prefrontal cortex during hot and cold reasoning. *NeuroImage, 20*(4), 2314–2321.

8. Richeson, J. A., Baird, A. A., Gordon, H. L., Heatherton, T. F., Wyland, C. L., Trawalter S., et al. (2003). An fMRI investigation of the impact of interracial contact on executive function. *Nature Neuroscience, 12,* 1323–1328.

9. Jost, J. T., Glaser, J., Kruglanski, A. W., & Sulloway, F. J. (2003). Political conservatism as motivated social cognition. *Psychologcal Bulletin, 129*(3), 339–375. Kruglanski, A. W., & Jost, J. T. (2003, August 28). Political opinion, not pathology. *Washington Post*, p. A27.

10. Uchii, S. Kyoto University, Japan. (1991, September 24) . *Sherlock Holmes and probabilistic induction*. Paper presented at the Lunchtime Colloquium, Center for Philosophy of Science, University of Pittsburgh. www.bun .kyoto-u.ac.jp/~suchii/holmes_1.html.

11. Thanks to Science Help Online Chemistry. www.fordhamprep.org/ gcurran/sho/sho/lessons/lesson73.htm

12. Goel, V., & Dolan, R. J. (2000). Anatomical segregation of component processes in an inductive inference task. *Journal of Cognitive Neuroscience, 12*(1), 110–119. Goel, V., & Dolan, R. J. (2004). Differential involvement of left prefrontal cortex in inductive and deductive reasoning. *Cognition, 93*(3), B109–121.

13. Elliott, R., Dolan, R. J., & Frith. C. D. (2000). Dissociable functions in the medial and lateral orbitofrontal cortex: Evidence from human neuroimaging studies. *Cerebral Cortex, 10*(3), 308–317.

14. Arana, F. S., Parkinson, J. A., Hinton, E., Holland, A. J., Owen, A. M., & Roberts, A. C. (2003). Dissociable contributions of the human amygdala and orbitofrontal cortex to incentive motivation and goal selection. *Journal of Neuroscience, 23*(29), 9632–9638.

15. Goel, V., & Dolan, R. J. (2001). The functional anatomy of humor: Segregating cognitive and affective components. *Nature Neuroscience, 4*(3), 237–238.

16. Neisser, U. (September–October 1997). Rising scores on intelligence tests. *American Scientist*, 440–447. Spearman spelled it out at length in *The abilities of man* (New York: Macmillan, 1927).

17. Duncan, J. (2003). Intelligence tests predict brain response to demanding task events. *Nature Neuroscience, 6*(3), 207–208.

18. The answer is set 3: The letters are equally spaced in the alphabet. Duncan, J. (2000). A neural basis for general intelligence. *Science, 289*(5478), 457–460. See also Duncan, J. (2005). Frontal lobe function and general intelligence: Why it matters. *Cortex, 41*(2), 215–217.

19. Angier, Natalie. (2000, July 21). Study finds region of brain may be key problem solver. *New York Times*, p. A11.

20. Ibid.

21. Sternberg, R. J. (2000). Cognition: The holey grail of general intelligence. *Science, 289*(5478), 399–401.

22. Siegfried, Tom. (2002, January 14). Origin of intelligence differences is gray area. DallasNews.com, in reference to Plomin, R., & Kosslyn, S. M. (2001). Genes, brain and cognition. *Nature Neuroscience, 4*, 1153–1154.

23. Gray, J. R., Chabris, C. F., & Braver, T. S. (2003). Neural mechanisms of general fluid intelligence. *Nature Neuroscience, 6*(3), 316–322.

24. Neisser, U. Rising scores on intelligence tests. Studies in practical applica-

tions have found that "even the Raven's test generally results in a high degree of adverse impact on minority group and female applicants, which could lead to legal difficulties as well as perpetuate differences in social and economic status between ethnic groups in our society . . . Yet at the same time, in a puzzling fashion, Raven's and related gF tests, on average, have shown an increase of about 15 points, or one standard deviation per generation, around the world in the last decades." This is according to the New Zealand political scientist James R. Flynn.

25. Important in doing a Raven's task is discovering and holding in mind rules that govern variation among the entries in the boxes. More difficult matrix problems involve more rules. Thus, in order to solve difficult matrix problems, you must ascertain a rule and then maintain it while searching to discover a second rule, and so on. This ability to hold online goal-relevant rules in the face of parallel searches for new rules, distractions, and while filtering out irrelevant features is essential for acing Raven's matrices. Carpenter, P. A., Just, M. A., & Shell, P. (1990). What one intelligence test measures: A theoretical account of the processing in the Raven Progressive Matrices Test. *Psychological Review, 97*(3), 404–431.

26. Conway, A. R., Cowan, N., & Bunting, M. F. (2001).The cocktail party phenomenon revisited: The importance of working memory capacity. *Psychonomic Bulletin & Review, 8*(2), 331–335.

27. Goldinger, S. D., Kleider, H. M., Azumka, T., & Beike, D. R. (2003). "Blaming the victim" under memory load. *Psychological Science, 14*(1), 81–85.

28. Braver, T. S., & Bongiolatti, S. R. (2002). The role of frontopolar cortex in subgoal processing during working memory. *NeuroImage, 15*(3), 523–536.

29. Christoff, K., & Gabrieli, J. D. (2000). The frontopolar cortex and human cognition: Evidence for a rostrocaudal hierarchical organization within the human prefrontal cortex. *Psychobiology, 28*(2), 168–186. Christoff, K., et al. (2001). Rostrolateral prefrontal cortex involvement in relational integration during reasoning. *NeuroImage, 14*, 1136–1149.

30. The neuroscientist Joseph LeDoux, well-known for his studies of emotional circuits, suggests that implicit and explicit aspects of the self overlap, but not completely. See LeDoux, Joseph (2002). *Synaptic self: How our brains become who we are*. New York: Viking.

31. Christoff, K., Ream, J. M., & Gabrieli, J. D. (2004). Neural basis of spontaneous thought processes. *Cortex, 40*(4–5), 623–630.

3. Passion: In Cold Blood?

1. Burton, Robert. (1621). *The anatomy of melancholy*, section II, member III, subsection III—Division of Perturbations. www.psyplexus.com/burton/10.htm.

2. Anderson, S. W., Bechara, A., Damasio, H., Tranel, D., & Damasio, A. R. (1999). Impairment of social and moral behavior related to early damage in human prefrontal cortex. *Nature Neuroscience, 2*(11), 1032.

3. Mathieson, B. B., Forster, P. L., & Svendsen H. A. (2004). Affect regulation and loss of initiative in a case of orbitofrontal injury. *Neuro-Psychoanalysis*, 6(1), 47–62.

4. Aristotle (350 B.C.E.). *Rhetoric*, book II, chapter 1. Translated by W. Rhys Roberts, 1954. http://classics.mit.edu/Aristotle/rhetoric.2.ii.html.

5. Canli, T., & Amin, Z. (2002). Neuroimaging of emotion and personality: Scientific evidence and ethical considerations. *Brain and Cognition*, 50(3), 414–431. The orbitofrontal cortex, for instance, is the subject of massive scrutiny. The Vanderbilt neuropsychologist David Zald is editing an academic volume on the OFC, and already the text is over six hundred pages long. It would have been easier to write a decade ago, he says.

6. Barbas, H., Saha, S., Rempel-Clower, N., & Ghashghaei, T. (2003). Serial pathways from primate prefrontal cortex to autonomic areas may influence emotional expression. *BMC Neuroscience*, 4(1), 25.

7. Flaherty, Alice W. (2004). *The midnight disease*. New York: Houghton Mifflin.

8. Davidson, R. J. (2003). Seven sins in the study of emotion: Correctives from affective neuroscience. *Brain and Cognition*, 52(1), 129–132.

9. Jaynes, Julian. (1976). *The origin of consciousness in the breakdown of the bicameral mind*. New York: Houghton Mifflin. This book was a catalyst for much popular thinking about hemispheric divisions.

10. Ueda K., Okamoto, Y., Okada, G., Yamashita, H., Hori, T., & Yamawaki, S. (2003). Brain activity during expectancy of emotional stimuli: An fMRI study. *NeuroReport*, 14(1), 51–55.

11. Kawasaki, H., Kaufman, O., Damasio, H., Damasio, A. R., Granner, M., Bakken, H., et al. (2001). Single-neuron responses to emotional visual stimuli recorded in human ventral prefrontal cortex. *Nature Neuroscience*, 4(1), 15–16.

12. See Queendom.com. Hamilton, Anita. (2002, September 30). What breed of dog are you? *Time*. Online at www.time.com/time/archive/preview/0,10987,353587,00.html.

13. Davidson, R. J., Jackson, D. C., & Kalin, N. (2000). Emotion, plasticity, context, and regulation: Perspectives from affective neuroscience. *Psychological Bulletin*, 126(6), 890–909.

14. Davidson, R. J. (2002). Anxiety and affective style: Role of prefrontal cortex and amygdala. *Biological Psychiatry*, 51(1), 68–80.

15. Zald, D. H., Mattson, D. L., & Pardo, J. V. (2002). Brain activity in ventromedial prefrontal cortex correlates with individual differences in negative affect. *Proceedings of the National Academy of Sciences*, 99(4), 2450–2454. This ventromedial landmark is also considered part of the cingulate cortex called the subgenual cingulate, and is located "under and behind" the anterior cingulated, which is associated more with cognitive operations.

16. Weiner, Ellis, & Chast, Roz. (2004). *The joy of worry*. San Francisco: Chronicle. Personality types react differently to addictive drugs as well. Psychiatrists at UC Irvine observed that hostile individuals with hair-trigger tempers may be more prone to nicotine addiction and have more trouble

kicking the cigarette habit than people with low-hostility profiles. Volunteers who showed higher tendencies to anger, aggression, impatience, and anxiety needed a higher dose of nicotine to get the same buzz as volunteers with higher happiness, relaxation, and curiosity profiles. (Type A personalities are more likely to be prodigious smokers than laid-back people, especially when agitated.) Numerous labs have scanned smokers' brains infused with nicotine to see what areas the drug targets. But the Irvine study added nonsmokers to the mix. In people more easily provoked to anger or agitation, the nicotine not only triggered dramatic changes in brain activity, but despite the common assumption that nicotine is calming, at higher doses it actually made them more aggressive. Principal investigator Steven Potkin called this the "born to smoke" brain response, suggesting that people with hostile, aggressive personality traits have a predisposition to nicotine addiction without ever lighting up a stick.

17. Starting with around $2,000, players selected cards from one of four decks on the computer screen. Each draw made them either richer or poorer. The healthy player soon learned that the left two decks were more likely to yield gains, but even heftier losses, and that the two right-hand decks yielded smaller but more frequent gains, with less chance of big losses. These players eventually began drawing cards only from the right-hand decks. Not so the patients.

18. Morgan, C. A, Wang, S., Rasmusson, A., Hazlett, G., Anderson, G., & Charney, D. S. (2001). Relationship among plasma cortisol, catecholamines, neuropeptide Y, and human performance during exposure to uncontrollable stress. *Psychosomatic Medicine, 63*, 412–422. Morgan, C. A., Rasmusson, A. M., Wang, S., Hoyt, G., Hauger, R. L., & Hazlett, G. (2002). Neuropeptide-Y, cortisol, and subjective distress in humans exposed to acute stress: Replication and extension of previous report. *Biological Psychiatry, 52*(2), 136–142.

19. Karama, S., Lecours, A. R., Leroux, J. M., Bourgouin, P., Beaudoin, G., Joubert, S., et al. (2002). Areas of brain activation in males and females during viewing of erotic film excerpts. *Human Brain Mapping, 16*(1), 1–13.

20. Gur, R.C., Gunning-Dixon, F., Bilker, W. B., & Gur, R.E. (2002). Sex differences in temporo-limbic and frontal brain volumes of healthy adults. *Cerebral Cortex, 12*(9), 998–1003.

21. To measure obedience to authority, Milgram famously persuaded people to punish others with electric shocks of increasing intensity, up unto "death." A majority of the volunteers did so, and with alacrity, zealously adhering to a protocol based on a set of "rules." Unbeknownst to the punishers, the shocks were faked, and the pain reactions of the "tortured" victims were staged. But the point was made.

22. Davidson, R. J., Marshall, J. R., Tomarken, A. J., & Henriques, J. B. (2000). While a phobic waits: Regional brain electrical and autonomic activity in social phobics during anticipation of public speaking. *Biological Psychiatry, 47*(2), 85–95.

23. See Litchtenstein Creative Media, www.lcmedia.com/mind251.htm; Jim Cramer's TheStreet.com; and Cramer, James J. (2002). *Confessions of a street addict.* New York: Simon & Schuster.

24. Harmon-Jones, E., Sigelman, J. D., Bohlig, J., & Harmon-Jones, C. (2003). Anger, coping, and frontal cortical activity: The effect of coping potential on anger-induced left frontal activity. *Cognition and Emotion, 17,* 1–24.

25. Pelletier, M., Bouthillier, A., Levesque, J., Carrier, S., Breault, C., Paquette, V., et al. (2003). Separate neural circuits for primary emotions? Brain activity during self-induced sadness and happiness in professional actors. *NeuroReport, 14*(8), 1111–1116.

26. Maney, Kevin. (2004, March 10). Money can't buy happiness, but happiness may buy money. *USA Today.* www.usatoday.com/tech/columnist/kevinmaney/2004-03-10-money_x.htm.

27. Ibid.

29. Markowitsch, H. J., Vandekerckhovel, M. M., Lanfermann, H., & Russ, M. O. (2003). Engagement of lateral and medial prefrontal areas in the ecphory of sad and happy autobiographical memories. *Cortex, 39*(4–5), 643–665.

29. Camille, N., Coricelli. G., Sallet, J., Pradat-Diehl, P., Duhamel, J. R., & Sirigu, A. (2004). The involvement of the orbitofrontal cortex in the experience of regret. *Science, 304*(5674), 1167–1170.

30. The study included three subjects with PFC lesions that mainly spared the OFC. These patients, like the normal participants, experienced regret when they made the wrong choices and lost money, thus further implicating the OFC as the mediator of this emotion.

31. Thompson, Clive. (2003, October 26). There's a sucker born in every medial prefrontal cortex. *New York Times.* BrightHouse Institute for Thought Sciences (Atlanta, GA, at Emory University) has become the first neuromarketing firm to be hired by a Fortune 500 consumer products company and is being roundly criticized for it. See Wahlberg, David (2004, February 1). Advertisers probe brains, raise fears. *Atlanta Journal-Constitution*, p. 2.

32. Berridge, K. C. (2003). Pleasures of the brain. *Brain and Cognition, 52*(1), 106–128. Berridge, K. C., & Robinson, T. E. (2003). Parsing reward. *Trends in Neuroscience, 26*(9), 507–513.

33. The shell of the nucleus accumbens also connects to primitive structures deep within the brain stem, like the parabrachial nucleus. The parabrachial nucleus modulates pleasurable sensations to gustatory and other bodily systems and in turn, send signals back up to the NAC and higher brain regions via the insula, which processes taste sensations and related emotions and thoughts related to the entire "selfness" of the body.

34. Small, D. M., Zatorre, R., Dagher, A., Evans, A. C., & Jones-Gotman, M. (2001). Changes in brain activity related to eating chocolate: from pleasure to aversion. *Brain, 124*(9), 1720–1733.

35. Bartels, A., & Zeki, S. (2000). The neural basis of romantic love. *NeuroReport, 11*(17), 3829–3834.
36. Wen, Patricia. (2001, February 14). In science, love now has a reality check. *Boston Globe*. www.boston.com/dailyglobe2/045/nation/In_science_love_now_has_a_reality_checkP.shtml.
37. Bartels, A., & Zeki, S. (2004). The neural correlates of maternal and romantic love. *NeuroImage, 21*(3), 1155–1166.
38. Levesque, J., Eugene, F., Joanette, Y., Paquette, V., Mensour, B., Beaudoin, G., et al. (2003). Neural circuitry underlying voluntary suppression of sadness. *Biological Psychiatry, 53*(6), 502–510.
39. Davidson, Richard J., & Harrington, Anne (Ed.) (2002). *Visions of compassion, Western scientists and Tibetan Buddhists examine human nature*. New York: Oxford University Press.
40. Tenzin Gyatso, the 14th Dalai Lama. (2003, April 26). The monk in the lab. Op-Ed, *New York Times*.
41. Davidson, R. J., Kabat-Zinn, J., Schumacher, J., Rosenkranz, M., Muller, D., Santorelli, S. F., et al. (2003). Alterations in brain and immune function produced by mindfulness meditation. *Psychosomatic Medicine, 65*(4), 564–570. The study took five years to publish in part because several higher-profile journals to which it was submitted refused even to send it out for peer review, according to Hall, Stephen S. (2003, September 14), Is Buddhism good for your health? *New York Times*.
42. www.halfbakery.com/idea/Fear_20Control_20Helmet.
43. de Quervain, D. J., Fischbacher, U., Treyer, V., Schellhammer, M., Schnyder, U., Buck, A., et al. (2004). The neural basis of altruistic punishment. *Science, 305*(5688):1254–1258.
44. Lorenz, J., Minoshima, S., & Casey, K. L. (2003). Keeping pain out of mind: The role of the dorsolateral prefrontal cortex in pain modulation. *Brain, 126*(5), 1079–1091.
45. Wager, T. D., Rilling, J. K., Smith, E. E., Sokolik, A., Casey, K. L., Davidson, R. J., et al. (2004). Placebo-induced changes in MRI in the anticipation and experience of pain. *Science, 303*(5661), 1162–1167.
46. Ochsner, K. N., Bunge, S. A., Gross, J. J., & Gabrieli, J. D. (2002). Rethinking feelings: An MRI study of the cognitive regulation of emotion. *Journal of Cognitive Neuroscience, 14*(8), 1215–1229.
47. Ibid.
48. In the acknowledgments to his novel *Saturday*, Ian McEwan thanks Dolan for his input, calling him "that most literary of scientists."
49. Gray, J. R. (2001). Emotional modulation of cognitive control: Approach-withdrawal states double-dissociate spatial from verbal two-back task performance. *Journal of Experimental Psychology: General, 130*, 436–452.
50. Gray, J. R., Braver, T. S., & Raichle, M. E. (2002). Integration of emotion and cognition in the lateral prefrontal cortex. *Proceedings of the National Academy of Sciences of the United States of America, 99*, 4115–4120.
51. Peyton-Dahlberg, Carrie. (2004, February 7). Brain research tapped to tell

what buyers like scientists, sellers exploring together. *Sacramento Bee.* www.commercialalert.org/news/featured-in/2004/02/brain-research-tapped-to-tell-what-buyers-like-scientists-sellers-exploring-together.

52. Wells, Melanie. (2003, September 1). In search of the buy button. *Forbes.* www.forbes.com/Forbes/2003/0901/062-print.html.

53. Tierney, John. (2004, April 20). Using M.R.I.'s to see politics on the brain. *New York Times*, p. A1.

54. Elias, Paul. (2004, October 28). Brain scanners can probe your politics. Associated Press. www.msnbc.msn.com/id/6356637.

4. Violence: Morality and the Minds of the Killers

1. Kubrick, S. A *Clockwork Orange*, www.screentalk.biz/moviescripts/A%20Clockwork%20Orange.pdf.

2. Anderson, S. W., Bechara, A., Damasio, H., Tranel, D., & Damasio, A. R. (1999). Impairment of social and moral behavior related to early damage in human prefrontal cortex. *Nature Neuroscience, 2*(11), 1032–1037.

3. Adolphs, R. (2003). Cognitive neuroscience of human social behaviour. *Nature Reviews Neuroscience, 4*(3), 165–178.

4. Hadjikhani, N., & de Gelder, B. (2003). Seeing fearful body expressions activates the fusiform cortex and amygdala. *Current Biology, 13*(24), 2201–2205. Landau, Misia. (2004, January 23). Brain takes similar approach to bodily, facial expressions. *Harvard Focus.* http://focus.hms.harvard.edu/2004/Jan23_2004/imaging.html.

5. Hasson, U., Nir, Y., Levy, I., Fuhrmann, G., & Malach, R. (2004). Inter-subject synchronization of cortical activity during natural vision. *Science, 303*(5664), 1634–1640.

6. Pesson, Luiz. (2004). Seeing the world in the same way. *Science, 303* (5664), 1618.

7. Singer, T., Seymour, B., O'Doherty, J., Kaube, H., Dolan, R. J., & Frith, C. D. (2004). Empathy for pain involves the affective but not sensory components of pain. *Science, 303*(5661), 1157–1162.

8. McGivern, R. F., Andersen, J., Byrd, D., Mutter, K. L., & Reilly, J. (2002). Cognitive efficiency on a match to sample task decreases at the onset of puberty in children. *Brain and Cognition, 50*(1), 73–89.

9. Hornak, J., Bramham, J., Rolls, E. T., Morris, R. G., O'Doherty, J., Bullock, P. R., et al. (2003). Changes in emotion after circumscribed surgical lesions of the orbitofrontal and cingulate cortices. *Brain, 126*(7), 1691–1712.

10. Shamay-Tsoory, S. G., Tomer, R., & Aharon-Peretz, J. (2005). The neuroanatomical basis of understanding sarcasm and its relationship to social cognition. *Neuropsychology, 19*(3), 288–300.

11. How Theory of Mind arises in the evolution of more complex brains, and its neural bases, are subject to much debate. There is vigorous contention, for example, about the degree to which great apes can infer the minds of others. It is difficult to test with any degree of certainty.

12. Rilling, J. K., Sanfey, A. G., Aronson, J. A., Nystrom, L. E., & Cohen, J. D. (2004). The neural correlates of theory of mind within interpersonal interactions. *NeuroImage, 22*(4), 1694–1703. In the Ultimatum game, when players were pitted against computers, the exchanges were largely ineffective at engaging these brain areas. But in the Prisoner's Dilemma, the computer exchanges, too, were able to excite this ToM network, albeit to a lesser extent than with human partners. So in Prisoner's Dilemma, perhaps the ToM brain system can also be turned on by thinking about the unobservable states of machine "minds," or possibly players imbued their computer partners with human attributes.

13. Ochsner, K. N., Knierim, K., Ludlow, D. H., Hanelin, J., Ramachandran, T., Glover, G., et al. (2004). Reflecting upon feelings: An fMRI study of neural systems supporting the attribution of emotion to self and other. *Journal of Cognitive Neuroscence, 16*(10), 1746–1772.

14. Colman, A. M. (2003). Depth of strategic reasoning in games. *Trends in Cognitive Sciences, 7,* 2–4.

15. Hedden, T., & Zhang, J. (2002). What do you think I think you think?: Strategic reasoning in matrix games. *Cognition, 855*(1), 1–36.

16. Morton, Carol C. (2002, May 3). Winners and losers exhibit model fighting behavior. *Harvard Focus.* http://focus.hms.harvard.edu/2002/May3_2002/research_briefs.html; see also *Harvard Focus,* Nov. 30, 2001.

17. Gibbons, Ann. (2004). Chimpanzee gang warfare. *Science, 304*(5672), 818–819.

18. Bookman, Jay. (2001, May 20). Expanding your mind: Magnetic mind reader watches as you think. *Atlanta Journal-Constitution,* p. B1.

19. Lewis, Dorothy Otnow. (1998). *Guilty by reason of insanity: A psychiatrist explores the minds of killers.* New York: Ivy Books, 321.

20. Pincus, Jonathan H. (2001). *Base instincts: What makes killers kill?* New York: W.W. Norton, 78.

21. Filley, C. M., Price, B. H., Nell, V., Antoinette, T., Morgan, A. S., Bresnahan, J. F., et al. (2001).Toward an understanding of violence: Neurobehavioral aspects of unwarranted physical aggression: Aspen Neurobehavioral Conference consensus statement. *Neuropsychiatry, Neuropsychology, and Behavioral Neurology, 14*(1), 1–14.

22. Brower, M. C., & Price, B. H. (2001). Neuropsychiatry of frontal lobe dysfunction in violent and criminal behaviour: A critical review. *Journal of Neurology, Neurosurgery, and Psychiatry, 71*(6), 720–726.

23. Bergvall, A. H., Wessely, H., Forsman, A., & Hansen, S. (2001). A deficit in attentional set-shifting of violent offenders. *Psychological Medicine, 31*(6), 1095–1105.

24. Raine, A., Buchsbaum, M., & LaCasse, L. (1997). *Journal of Biological Psychiatry, 42*(6), 495–508.

25. Foreman, Judy. (2002, April 29). Routs of violence may lie in damaged brain cells. *Los Angles Times,* p. 1.

26. Hare, R. D. (September 1995). Psychopaths: New trends in research, the

Harvard Mental Health Letter. Cited in Sabbatini, R. M. E., The psychopath's brain, in *Brain & Mind: Electronic Magazine on Neuroscience*. www.cerebromente.org.br/n07/downcas/index.html#introduction.

27. Raine, A., Meloy, J. R., Bihrle, S., Stoddard, J., LaCasse, L., & Buchsbaum, M. S. (1998). Reduced prefrontal and increased subcortical brain functioning assessed using positron emission tomography in predatory and affective murderers. *Behavioral Sciences and the Law, 16*(3), 319–332. Pincus, *Base Instincts*, 56.

28. Raine, A., Phil, D., Stoddard, J., Bihrle, S., & Buchsbaum, M. (1998). Prefrontal glucose deficits in murderers lacking psychosocial deprivation. *Neuropsychiatry, Neuropsychology, and Behavioral Neurology, 11*(1), 1–7.

29. Raine, Adrian. (January 2002). New Hot Paper Comments. www.esi-topics.com/nhp/comments/january-02-AdrianRaine.html

30. Suplee, Curt. (2000, February 15). In the brains of the violent, gray matter may matter. *Washington Post*, p. A2.

31. Bender, Eve. (2003). Understanding aggression: It's largely in the planning. *Psychiatric News, 38*(23). http://pn.psychiatryonline.org/cgi/content/full/38/23/21.

32. Stuss, Donald T., & Knight, Robert T. (2002). *Principles of frontal lobe function*. New York: Oxford University Press.

33. Moeller does *not* recommend you try tryptophan depletion at home. Since aggressive behavior after lowering serotonin varied among the men in the study varied, some individuals might be prone to this effect, such as those with a history of aggressive behavior in childhood. Such people, Moeller suggested, might benefit from the plethora of SSRI medications that increase serotonin. But it was not difficult to fantasize a *Clockwork Orange* scenario, in which nice folks go out to Ye Olde Ultraviolence Pub and guzzle low-trypo drinks, before going on to rampage the neighborhood.

34. Marsh, D. M., Dougherty, D. M., Moeller, F. G., Swann, A. C., & Spiga, R. (2002). Laboratory-measured aggressive behavior of women: Acute tryptophan depletion and augmentation. *Neuropsychopharmacology, 26*(5), 660–671.

35. Gibbons, Ann. (2004). Tracking the evolutionary history of a "warrior" gene. *Science, 304*(5672), 818.

36. Brunner, H. G., Nelen, M., Breakefield, X. O., Ropers, H. H., & van Oost, B. A. (1993). Abnormal behavior associated with a point mutation in the structural gene for monoamine oxidase A. *Science, 262*(5133), 578–580.

37. Caspi, A., McClay, J., Moffitt, T. E., Mill, J., Martin, J., Craig, I. W., et al. (2002). Role of genotype in the cycle of violence in maltreated children. *Science, 297*(5582), 851–854.

38. Fonagy, Peter. (1999, May 13). *Transgenerational consistencies of attachment: A new theory*. Paper presented to the Developmental and Psychoanalytic Discussion Group, American Psychoanalytic Association Meeting, Washington, DC. http://psychematters.com/papers/fonagy2.htm

39. Schore, Allan N. (2001). The effects of a secure attachment relationship on

right brain development, affect regulation, and infant mental health. *Infant Mental Health Journal, 22,* 7–66. http://trauma-pages.com/a/schore-2001a.php.

40. Lewis, *Guilty by reason of insanity,* 331.

41. Goldberg, Carey. (2004, June 1). Scientists watch the brain wrestle with moral dilemmas. *Boston Globe.* www.boston.com/news/globe/health_science/articles/2004/06/01/scientists_watch_the_brain_wrestle_with_moral_dilemmas.

42. Although he has a grant from the Department of Defense to look into an fMRI model of a lie detector, Langleben claims his instrumentation is far from ready for use in, say, a large population of convicts, or middle-level corporate executives. At present, the machine is unable to discriminate among the brain activities in an individual in a one-shot session. A laboratory card game, moreover, in no way resembles the complex prevarication of a pathological liar in a profession of crime or high-level government office. Should it ever become a forensic option, such machines may still be unusable under the Fifth Amendment—sanction against self-incrimination.

43. Rilling, J., Gutman, D., Zeh, T., Pagnoni, G., Berns, G., & Kilts, C. (2002). A neural basis for social cooperation. *Neuron, 35*(2), 395–405.

44. de Quervain, D. J., Fischbacher, U., Treyer, V., Schellhammer, M., Schnyder, U., Buck, A., et al. (2004). The neural basis of altruistic punishment. *Science, 305*(5688), 1254–1258.

45. Knutson, Brian. (2004). Sweet revenge? *Science, 305*(5688), 1246–1247.

46. Greene, J. D., Nystrom, L. E., Engell, A. D., Darley, J. M., & Cohen, J. D. (2004). The neural bases of cognitive conflict and control in moral judgment. *Neuron, 44*(2), 389–400.

47. Josh Greene adds, "The reaction time effect doesn't necessarily hold within dilemmas and may, instead, be generated more by differences between dilemmas. For example, in the Crying Baby case, the reaction times were about the same for the yes and no answers. However, there are some cases that are almost always quick and almost always "no" (e.g., you don't like your boss—is it okay to push him off a cliff?) Overall, "yes" ends up being faster than "no" because there are no dilemmas (involving personal moral violations) that get fast yes answers, and not necessarily because yes answers are slower than no answers for any given dilemma."

48. Greene, J. D., Sommerville, R. B., Nystrom, L. E., Darley, J. M., & Cohen, J. D. (2001). An fMRI investigation of emotional engagement in moral judgment. *Science, 293*(5537), 2105–2108.

49. Greene, J. (2003). From neural "is" to moral "ought": What are the moral implications of neuroscientific moral psychology? *Nature Reviews Neuroscience, 4*(10), 846–849.

5. Creativity: Art as a Window into the Brain

1. Zeki, S. (2001). Essays on science and society. Artistic creativity and the brain. *Science, 293*(5527), 51–52.

2. Shimamura, A. P. (2002). Muybridge in motion: Travels in art, psychology and neurology. *History of Photography, 26*, 341–350.

3. University of Toronto Press release via 2003 ScienceDaily.com December 1. www.sciencedaily.com/releases/2003/10/031001061055.htm Carson, S. H., Peterson, J. B., & Higgins, D. M. (2003). Decreased latent inhibition is associated with increased creative achievement in high-functioning individuals. *Journal of Personality and Social Psychology, 85*(3), 499–506.

4. Ibid.

5. Miller, B. L., Boone, K., Cummings, J. L., Read, S. L., & Mishkin, F. (2000). Functional correlates of musical and visual ability in frontotemporal dementia. *British Journal of Psychiatry, 176*, 458–463.

6. Chamberlain, Claudine. (2000, October 20). An artful madness: Talents emerge, dementia takes over. ABCNews.com.

7. Nowakowska, C., Strong, C. M., Santosa, C. M., Wang, P. W., & Ketter, T. A. (2005). Temperamental commonalities and differences in euthymic mood disorder patients, creative controls, and healthy controls. *Journal of Affective Disorders, 85*(1–2), 207–215. Stanford news release. (2002, May 21). http://mednews.stanford.edu/news_releases_html/2002/mayreleases/creative_gen.html.

8. Seger, C. A., Poldrack, R. A., Prabhakaran, V., Zhao, M., Glover, G. H., & Gabrieli, J. D. (2000). Hemispheric asymmetries and individual differences in visual concept learning as measured by functional MRI. *Neuropsychologia, 38*, 1316–1324.

9. Sternberg, Robert J. (Ed.). (1999). *Handbook of creativity.* Cambridge, England: Cambridge University Press, 145.

10. Seger, C. A., Desmond, J. E., Glover, G. H., & Gabrieli, J. D. (2000). Functional magnetic resonance imaging evidence for right-hemisphere involvement in processing unusual semantic relationships. *Neuropsychology, 14*(3), 361–369.

11. Ilari, B., Polka, L., & Costa-Giomi, E. (2002, June 6). *Babies can unravel complex music.* Paper presented at the 143rd Meeting of the Acoustical Society of America. www.acoustics.org/press/143rd/Ilari.html.

12. Ohnishi, T., Matsuda, H., Asada, T., Aruga, M., Hirakata, M., Nishikawa, M., et al. (2001). Functional anatomy of musical perception in musicians. *Cerebral Cortex, 11*(8), 754–760.

13. Overy, K., Norton, A. C., Cronin, K. T., Gaab, N., Alsop, D. C., Winner, E., et al. (2004). Imaging melody and rhythm processing in young children. *NeuroReport, 15*(11), 1723–1726. The story of a melody "just coming" under duress: Samuel Barber gets a call from the pianist John Browning, saying, "Sam, it's less than two weeks before the premiere of the piano concerto, and I haven't got the third movement yet." Barber replies, "That's because I can't seem to think of a theme for it yet!" A day or two later, according to Browning, Barber is looking at himself in the mirror while shaving, and lo and behold, a melody presents itself. Barber puts down the razor, jots down the melody, and writes the whole movement in a few days.

14. Janata, P., Birk, J. L., Van Horn, J. D., Leman, M., Tillmann, B., & Bharucha, J. J. (2002). The cortical topography of tonal structures underlying Western music. *Science, 298*(5601), 2167–2170.
15. Duffles, Marilia. (2002, March 2). Secrets of human thinking. *Financial Times*. http://cogweb.ucla.edu/steen/2002-03-02_Neuroesthetics.html.
16. Cela-Conde, C. J., Marty, G., Maestu, F., Ortiz, T., Munar, E., Fernandez, A., et al. (2004). Activation of the prefrontal cortex in the human visual aesthetic perception. *Proceedings of the National Academy of Sciences of the United States of America, 101*(16), 6321–6325.
17. Zeki, Semir. (1999). *Inner vision: An exploration of art and the brain*. Oxford: Oxford University Press, 26.
18. Epstein, R. (2004). Consciousness, art, and the brain: Lessons from Marcel Proust. *Consciousness and Cognition, 13*(2), 213–240.
19. Mangan, B. (2001). Sensation's ghost: The non-sensory "fringe" of consciousness. *Psyche, 7*(18). http://psyche.cs.monash.edu.au/v7/psyche-7-18-mangan.html.

6. Silicon Minds: The Rise of Machine Genius

1. Fowler, Jonathan. (2004, October 20). U.N. predicts boom in robot labor. CBS News.com. www.cbsnews.com/stories/2004/10/20/tech/main650274.shtml.
2. Duke News & Communications. (2004, July 14). News tip: Films such as "I, Robot" affirm human superiority. www.dukenews.duke.edu/2004/07/robot_0704.html.
3. Miller, E. K., & Cohen, J. D. (2001). An integrative theory of prefrontal cortex function. *Annual Review of Neuroscience, 24*, 167–202.
4. Frank, M. J., Loughry. B., & O'Reilly, R. C. (2001). Interactions between frontal cortex and basal ganglia in working memory: A computational model. *Cognitive, Affective and Behavioral Neuroscience, 1*(2), 137–160. O'Reilly, R. C., & Frank, M. J., (2006). Making working memory work: Computational model of learning in the prefrontal cortex and basal ganglia. *Neural Computation, 18*(2), 283–328.
5. Frank, M. J., Seeberger, L. C., & O'Reilly, R. C. (2004). By carrot or by stick: Cognitive reinforcement learning in Parkinsonism. *Science, 306*(5703), 1940–1943.
6. In May 2005, Cornell University scientists announced that they had created small robots capable of building copies of themselves. Each robot consists of several four-inch cubes with identical machinery, electromagnets to attach and detach to each other, and a program for replication. The robots can bend and pick up and stack the cubes. "Although the machines we have created are still simple compared with biological self-reproduction, they demonstrate that mechanical self-reproduction is possible and not unique to biology," Hod Lipson said. Zykov, V., Mytilinaios, E., Adams, B., & Lipson, H. (2005). Self-reproducing machines. *Nature, 435*(7038), 163–164.

Index